Prediction of Concrete Durability

Prediction of Concrete Durability

Proceedings of STATS 21st Anniversary Conference

The Geological Society
London, UK
16 November 1995

Edited by

J. Glanville and A. Neville

CRC Press is an imprint of the
Taylor & Francis Group, an **informa** business
A CHAPMAN & HALL BOOK

CRC Press
Taylor & Francis Group
6000 Broken Sound Parkway NW, Suite 300
Boca Raton, FL 33487-2742

First issued in paperback 2019

ISBN-13: 978-0-419-21170-9 (hbk)
ISBN-13: 978-0-367-44821-9 (pbk)

Typeset in 10/12pt Palatino by Cambrian Typesetters, Frimley, Surrey

A Catalogue record for this book is available from the British Library

Visit the Taylor & Francis Web site at
http://www.taylorandfrancis.com

and the CRC Press Web site at
http://www.crcpress.com

Contents

Contributors

N.R. Buenfeld
Department of Civil Engineering
Imperial College of Science, Technology and Medicine
London

J.H. Bungey
Department of Civil Engineering
University of Liverpool

P.G. Fookes
Consultant
Winchester

J. Glanville
Glanville-STATS
St Albans

J. Newman
STATS
St Albans and Imperial College of Science, Technology and Medicine
London

G. Somerville
British Cement Association
Crowthorne

J.G.M. Wood
Structural Studies & Design Ltd
Chiddingfold

Foreword

The purpose of the foreword to a book is usually to extol it, to tell the readers how much the writer of the foreword agrees with the content of the book, and generally to commend it. In the case of this book, while I am willing to do all of the above, the situation is somewhat different.

First of all, this book is a collection of six papers, each by a different author, presented at a conference held in London on 16 November 1995. The papers are entirely self-standing but they are united by the theme of the performance prediction of concrete. Their coverage ranges from the consideration of aggregate, through modelling the transport phenomena in concrete and non-destructive assessment of expected performance, on to consideration of service life and, hopefully, to achievement of durable concrete. The scene for these papers is set by the first paper, which reviews the landmarks in the performance prediction in the years 1974 to 1995.

This period of 21 years was chosen because 1995 is the 21st anniversary of STATS, a firm of consulting engineers and scientists, located in St Albans, UK, headed by John Glanville. This anniversary genesis is the second special feature of the present book, and it merits some further comment. The name Glanville, *père et fils*, is closely associated with the progress of concrete technology in the UK. The contribution of Sir William Glanville in the post-war years cannot be overestimated, and it is appropriate that John Glanville, jointly with John Newman, opens the book with a review of changes in concrete practice over the last 21 years. It is interesting to see the parallel changes in the activity of STATS over the same period.

Consideration of the two concurrent developments makes me realize that materials consultancy can be as much a leader in development as investigations undertaken in a research laboratory. We expect academic research to lead to improved concrete (which it does not always achieve) but, all too often, we forget the contribution of those practitioners who by ingenious testing and by a more broadly-based approach to problems make it possible to improve concrete in the future. We should recognize the value of such work and encourage a more widespread transfer of findings by those involved in testing to concrete design and practice.

The theme of the conference, and therefore the title of the book, is

Prediction of Concrete Durability. The performance of concrete depends not only on its physical and mechanical properties at the time it is put into service but also on its subsequent response to the environment. This response over a period of time is called durability. Good durability of concrete was, in the past, taken for granted, all attention being concentrated on strength. However, numerous cases of deterioration, often very serious, of concrete in service have shifted much of the attention to ensuring durable concrete, at the same time recognizing the need for routine maintenance. We have abandoned the old claim of concrete as a maintenance-free material.

At this point I would like to sound a word of caution. We have to be careful not to lose sight of the continuing need for an appropriate strength as well as for the required mechanical and physical properties of concrete. It would be a mistake to replace overemphasis on strength by overemphasis on durability: both strength and durability have to be considered explicitly at the design stage.

The situation is then that we now recognize the very important criterion of durabilty of concrete in the design of structures, but we still fall short of being able to design explicitly for a required durability. This is a vital aim, and the chapters in this book contribute to the achievement of that aim. In consequence, I am able, as I said in the opening words of this foreword, to commend this book to all those involved in design and construction, in supervision and execution and in materials processing and supply, as well as in testing and in, regrettably, necessary repair work.

Finally, I would like to congratulate STATS on its celebration of 21 years of successful contribution to the world of better concrete. Many happy returns!

Adam Neville
London, November 1995

Acknowledgement

The editors wish to thank Cynthia Bloomfield for the very considerable effort involved in organizing the conference that provided the platform for the papers which form the content of this book.

With just the right combination of charm and determination she managed to obtain the contributions on time and, after enormous effort, to transcribe and sort out the discussion which always forms a most valuable part of a publication of this type.

Introduction and welcome by John Glanville

I want to welcome our Chairman, our speakers, our staff and everybody who has joined us today to celebrate our 21st birthday. Of course, it is the custom nowadays to celebrate 18th birthdays but we thought we would be traditional. STATS was formed by myself and my partner Geoff Gregg in 1974 and I am sure he recalls the day as well as I do. My father, Sir William Glanville, was our first consultant and I well remember his interest and enthusiasm in our new venture. I should particularly like to thank Dr Adam Neville for acting as our Chairman and all the speakers who have spent an enormous amount of time and energy in preparing their papers, the presentation of which we eagerly await. In addition I should particularly like to thank my colleague, Cynthia Bloomfield; she has organized the whole thing single-handed with energy and efficiency.

As you know, the proceedings are to be published by Spon and they will print not only the papers but also the discussion; the price is fixed so the more you discuss the more they will have to print and the better value the book will be.

STATS and I have had a love–hate relationship for 21 years. I love the challenge but hate the commercial reality which has always tended to make profit so elusive. STATS was a hobby for many years until commercial reality took over; today it is great fun but the commercial climate is, of course, very worrying for everybody connected with construction.

As I have said, we are extremely fortunate to have Dr Neville as our Chairman today. He is known to all of us as the doyen of reinforced concrete: an academic, author and expert consultant of international renown. Few engineers have the ability to explain complex matters in a simple way; Dr Neville is one of them and *Properties of Concrete*, the best known of his books, has, since 1963, helped practising engineers to gain knowledge of the properties and behaviour of concrete.

After a distinguished academic and professional career and at an age when most men have retired from active life, Dr Neville is as busy as ever, not only professionally but also on the ski slopes whenever he gets the chance.

In introducing the proceedings of a symposium held by Canmet/ACI in June this year to honour Dr Neville, Mohan Malhotra wrote, ‘this publication is a highly deserved tribute to his many outstanding and distinguished contributions in the general area of concrete technology and civil engineering’.

Ladies and gentlemen, please welcome our Chairman – Dr Adam Neville.

Opening remarks by the Chairman, Dr Adam Neville

Good morning ladies and gentlemen. John, thank you for your introduction. I am not sure I liked all your references to age; I am saying that because a few months ago I gave a lecture in Canada which was on a structural analysis topic, no reference to concrete, and after the lecture a young man came to me and said, 'Are you the Neville who wrote *Properties of Concrete*?' I said 'Yes'. Back came his answer: 'I thought you were dead.' Well, as you have heard, I am your Chairman; the Chairman's role is very limited – he doesn't contribute knowledge; all he has to do is to prevent more than two people speaking simultaneously. Otherwise, he just has to look beautiful. There is no difficulty in that respect. However, your programme provides for an introduction – a 15 minute introduction. I don't think I shall speak for 15 minutes but there are two things I would like to say.

One is to welcome you most warmly. There has been a tremendous response to the invitation sent out by John Glanville to this conference, and I think we have a very good turnout and we have an audience which covers a wide spectrum of the construction industry and consists of people who are really important and influential. In consequence, I hope we will get a good interaction and hopefully a good discussion and cross-fertilization. Well, if we do, that should lead to better concrete in the future and this is the purpose of the conference.

The second thing which I wish to say is to give the genesis of this conference slightly more fully than John Glanville gave us. It is a birthday celebration of STATS and you may wonder, as I did, about the origin of the name. I am told that the firm was originally called St Albans Testing Services and when it started its operation one of the people in the laboratory was asked to mark all the property of the company by painting on its name. The man chose the acronym S–T–A–T–S, so STATS came into being and is, I think, the registered name of the company. Well, you heard that STATS was formed by John Glanville and Geoff Gregg; they spent the early years investigating defects and failures in concrete structures. Later on, in the late 1970s, the expertise of STATS grew and came to cover a wider range of engineering and building matters, and by the end of the 1970s the firm employed chemists and geologists as well as engineers, all of whom specialized in investigation of problems.

Non-destructive testing, which looms large, and about which we shall have a paper, became a major investigating tool and it was employed on a highway bridge for the Department of Transport in 1979 when STATS was appointed to carry out site work. That is how it got into what John has described as a love–hate relationship with reinforced and prestressed concrete. It continued to grow in size and in the breadth of services offered and today it has a geotechnical department, environmental division, which undertakes contaminated site investigations, environmental audits and design of remedial schemes, and its material group carries out a wide range of investigations into building stone, clay, render, mortar, screed, brick, block, coating, roofing – and I don't know if there is anything else. It also has accredited laboratories for compliance and for investigations, and has inspectors in steel fabrication yards and precast concrete works.

Well, we are not going to deal with all that today and, despite the classical moaning about profits of consulting engineers, I have not found that I have to buy John a lunch and I think the firm is doing very well; the turnover is, I think, £4 million; he says they make a modest profit, but then he is a modest man.

I think all those present here are involved one way or another in the construction industry. We are proud to be involved but this attitude is not shared by the public at large who are mesmerized by electronics, by service industries and the City of London (I don't know about Barings) but who view construction as dirty work that despoils the countryside. Well, who is to blame for this? We are, for we are too modest about the vital contribution of construction to the quality of life. At the risk of stating the obvious, for the purpose of your reminding anybody with whom you have a drink later today or tomorrow who is not in construction, let me remind you that construction represents an annual expenditure of £50 billion. This includes new structures, buildings, maintenance, repair and refurbishment. Ten per cent of this, that is £5 billion, is spent on concrete; to put it another way, £100 per annum is spent, so to speak, by every man, woman and child, on concrete – probably as much as on beer.

Alas, a significant proportion of the construction expenditure covers repairs and much more is needed to be spent on repairs. For example, the Department of Transport, in its report on bridges in 1989, estimated £620 million then as being required for repairs. Well, what I am trying to say is that we must, I repeat, must do better in the future. It is therefore highly appropriate that the theme of this conference is the performance prediction of concrete. The performance of concrete depends not only on its physical and mechanical properties at the time when it is put into service, but also on the subsequent response of concrete to the environment. The response over a period of time is called durability. Good durability of concrete was taken for granted, all attention being concentrated on strength. However, recently, deterioration, often very serious, of concrete in service has shifted much of the attention to ensuring durable concrete, at the same time recognizing the need for routine maintenance. We have at last abandoned the old claim of concrete as a maintenance-free material. Having said that, I would like to sound a word of caution. We have to be careful not to lose sight of the continuing need for an appropriate strength as well as the required mechanical and physical properties of concrete. It would be a mistake to replace overemphasis on strength by overemphasis on durability. Both strength and durability have to be considered explicitly at the design stage. The situation is then that we recognize now the very important criterion of durability of concrete in the design of structures but we still fall short of being able to design explicitly for the required durability. This is a vital aim and the conference is going to help us to move in that direction. We must move, and move quickly, towards more durable structures and a prerequisite for this is performance prediction.

Let us then hear the papers helping us to achieve this aim. Thank you.

1

The performance prediction of concrete: review of landmarks 1974–95

by J. Glanville and J. Newman

ABSTRACT

Since the early 1930s vast strides have been made in the development of concrete as a material and of reinforced concrete as a structural material. Over the last 21 years steady progression and development has produced many improvements and some failures. The chapter charts these and the growth of STATS as an engineering materials consultancy.

Keywords: Cathodic protection, concrete, high-alumina cement, testing.

1.1 INTRODUCTION

In the period immediately following World War I much fundamental research was carried out into the properties of cement and concrete. John Glanville's father, W.H. Glanville, joined the staff of the Building Research Station (BRS) just before they moved to Garston in 1922 and was involved in the pioneering research undertaken primarily to assist in the preparation of the first code of practice for reinforced concrete, which was published in 1934 [1]. John Newman's father joined the BRS staff in the early 1930s and started to work on concrete materials, with particular reference to aggregates and mix design. His research was mainly related to improving the materials aspects of codes of practice, since the performance of concrete as a material had been of increasing concern to the engineering profession and the only existing research was of a relatively simple nature undertaken by practising engineers.

Such work was reported in a paper presented to the Institution of Civil Engineers in 1927 by Dr Oscar Faber on the plastic yield, shrinkage and other problems of concrete and their effect on design [2]. This reported experimental work carried out on four concrete beams in the laboratories run by Mr Stanger and the discussion included contributions from Professor Dixon (Professor of Civil Engineering at Imperial College), Sir Owen Williams, Professor Lee, Mr Haddon-Adams, Dr Glanville, Mr Scott (who subsequently founded Scott and Wilson), Professor Johnson of Baltimore USA, Mr Lord of Chicago, Professor Otto Graf of Stuttgart, Dr Lowe-Brown (who remarked that he had been undertaking research for more than 32 years and had presented his first paper to the Institution in 1897), Professor Morris of Columbus, Ohio, and Dr Unwin, President of the Institution, among others – such was the thirst for knowledge of concrete.

Shortly after the publication by the Department of Scientific and Industrial Research of the code of practice under the title *Recommendations for a Code of Practice for the Use of Reinforced Concrete in Buildings* (later to become CP114 [3]). John Glanville's father, who had been technical officer to the committee and was currently chief engineer of the BRS, and W.L. Scott produced the first handbook on the code which, after many revisions and additions, finished up in 1973 as *The Explanatory Handbook on the BS Code of Practice for Reinforced Concrete CP114, 1957* [4], with metric appendix and incorporating an amendment issued in February 1965 [5].

Throughout this period the codes had stated that the strength of Portland cement concrete increases appreciably with age. Table 2A of the 1973 publication showed concrete crushing strength as being about 40% greater after one year than at 28 days and the code permitted the use of an age factor whereby a factor of 1.24 could be applied to the permitted compressive stress in concrete after 12 months. Similar allowances were permitted under the replacement code CP110, *Code of Practice for the Structural Use of Concrete*, in 1972 [6], but they were subsequently withdrawn because of changes in the performance of Portland cement and do not form a feature of current design considerations. The use of calcium chloride to enhance the rate of hardening of concrete was permitted by CP114, which stated that 'not more than 2% by weight of the cement should be used'. In the explanatory handbook it said, 'an excessive amount of calcium chloride may cause too rapid setting, leading to difficulty in placing the concrete and may also result in corrosion of the reinforcement.' The use of calcium chloride was in fact finally banned by an amendment to CP110 in 1977.

Since concrete made with high-alumina cement gains strength much more quickly than that made with ordinary Portland cement, it is ideally suited in many respects to use in precast concrete work. According to

CP114, 'the use of high-alumina cement requires much greater care than the use of Portland type cement' and if it is not correctly treated during its early life or maintained in wet or humid conditions at temperatures above about 27°C conversion is likely to take place of the hydrated cement to a chemically stable, but porous, form. Table 31 gave permissible stresses for high-alumina cement concrete where conversion was likely to occur. The note in the explanatory handbook stated 'in recent years there has been considerable controversy about the extent of the danger of structural failure associated with the use of high-alumina cement.' Appendix B was added to the code in 1965 to give guidance on this matter [5] and read as follows: 'Provided proper precautions are taken, as set out in this clause, high-alumina cement concrete can be used with every prospect of its satisfactory behaviour throughout the life of the structure. The possibility of conversion can be reduced to negligible proportions if sufficient care is taken at the time of construction, so long as the structure will not subsequently be maintained in hot, moist conditions.' High-alumina cement was subsequently banned from use in 1975.

The state of the art in the early 1970s can be seen from the 1972 edition of Dr Adam Neville's seminal text book *Properties of Concrete* [7] which advises the reader that:

- the usual primary requirement of good concrete in its hardened state is a satisfactory compressive strength;
- the possibility of corrosion of reinforcing steel by calcium chloride is still a subject of some controversy, but the US Bureau of Reclamation – a very large user of concrete – found no indication that the use of calcium chloride, applied in correct proportions, results in corrosion of reinforcement, although calcium chloride has been found to lead to corrosion of prestressing wire and must not be used in the manufacture of prestressed concrete.

In addition:

- de-icing salt is noted as being liable to cause scaling of the concrete surface of road slabs but no reference is made to any adverse effects on reinforcing steel either in the road slabs themselves or in adjacent highway structures;
- the importance of the production of high-alumina cement concrete with a low water/cement ratio and with care being taken to avoid high temperatures is stressed. The conversion is fully explained and the loss of strength on conversion is documented, the conclusion being that, 'except for rich and strong mixes alumina cement concrete should not be generally used in structural members';

- problems associated with alkali–aggregate reactivity are considered in detail but the conclusion is reached that, 'the alkali aggregate reaction of the type described has fortunately not been encountered in Great Britain but is widespread in many other countries, notably North America, Scandinavia, India, Australia and New Zealand'.

In addition to these technical matters there was, in the period immediately following World War II, extreme pressure on the Government to meet targets for the construction of housing, roads, public buildings and other structures. At the same time, commercial development was gaining pace. This led to the building boom of the 1960s and early 1970s. There was a need to produce structures rapidly at the lowest possible cost, with inexperienced labour and, at the same time, to cope with major disruptions such as a serious shortage of reinforcement in the period around 1960, periodic shortages of cement, and an energy crisis in the early 1970s. These were coupled with a steady rise in the bank rate from 1950 to the mid-1970s. By 1974 when STATS was formed by John Glanville and his partner, Geoff Gregg, and Probe Technical Services (Probe) by John Newman and his colleagues, there was therefore pressure for the introduction of new techniques and technologies. The system-building boom of the 1960s had left a legacy of problems and there was, therefore, great demand for both routine testing and investigative engineering.

This sets the scene in 1974 for our entry into the construction materials consulting fraternity, a body then dominated by Messrs Sandberg and Stanger, long-established firms both of which date from the 19th century. As young men with energy and enthusiasm, we reckoned we were better than all the others and could show them how to apply proper scientific principles to construction materials' technology. Well, we have had our successes and failures and now, no longer being young men, are much more modest in our claims; others are now trying to show **us** the way.

As with many ideas which have been developing in the minds of their proponents for a number of years, a catalyst is necessary to create action. In our case the catalyst was high-alumina cement concrete. As has been noted above, it was common knowledge in the 1960s that high-alumina cement concrete would convert to a more stable but weaker form if it were not correctly treated in its early life, the effects of such conversion being worse for concretes with high water/cement ratios. In practice, the concretes were seldom made with low water/cement ratios and they were not always correctly treated, thus they did convert; as a result of this, a number of beams on the roof of a swimming pool at Sir John Cass Primary School collapsed. The confusion which followed got us going.

1.2 CONCRETE STRENGTH

As Dr Adam Neville succinctly put in the preface to his book [7], 'the usual primary requirement of a good concrete in its hardened state is compressive strength. This is aimed at not only so as to ensure that the concrete can withstand a prescribed compressive stress but also because many other desired properties of concrete are concomitant with high strength.' This statement is, of course, as correct today as when it was written; however, our understanding of some of the qualities of the concrete which are not directly associated with strength has increased considerably. In the early 1980s a working party of the Concrete Society considered the changes in the properties of ordinary Portland cement which had occurred and their effects on concrete [8]. In the introduction they stated, 'in recent years there has been a realisation by engineers that the cements they are using have changed gradually but significantly in some of their properties, the most noticeable change being an increase in strength. Concern has been experienced about the implications of these changes and their possible effect . . .' The report considers a great deal of evidence and concludes that whilst the total proportion of silicates in cement has remained about constant, the ratio of the tri-calcium silicate C_3S to the di-calcium silicate C_2S has risen and that this has resulted in a marked increase in the early strength of concrete made with a given quantity of cement, this strength increase being most noticeable at an early age.

These changes would appear to have been brought about by a change in manufacturing technique whereby, instead of using a wet slurry process, a dry grinding and pelletization system was introduced as new cement works were brought into production in the 1960s and 1970s. This early development of strength appeared to be of considerable benefit to engineers since it resulted in the reduction of the cost of concrete because less cement could produce the same strength as had previously been obtained by a richer mix and allowed higher water/cement ratios to be used to achieve a given strength. Little attention was paid by the average construction engineer to the influence of such changes on the durability of structures.

Since the principal preoccupation of the UK construction industry was the strength of the concrete, the cube crushing test was universally used for quality control and design purposes. The New Unified Code for Concrete (which became CP110) introduced in the early 1970s adopted a probabilistic approach to concrete strength. This forced engineers to be aware of the importance of the variability of concrete and the significant influence of the design of the machine for testing the cube on the measured strength of concrete. Until that time British Standards permitted the use of any testing machine provided the load indication was calibrated in accordance with BS 1610. However, awareness grew rapidly in the early 1970s of the need to design testing machines for such characteristics as stiffness and, in particular, spherical seating performance.

At that time laboratory accreditation schemes were unheard of in the UK. However, for obvious commercial reasons the ready-mixed concrete industry found the wide variation in test results unacceptable and this led to the improved specification of testing machinery and the introduction of the British Ready Mixed Concrete Association accreditation scheme. This was in due course absorbed into the government's NATLAS (now termed NAMAS) scheme and we were among the first laboratories to join what is now a universally accepted accreditation system.

As a sidelight it is interesting to remember that in the early 1970s slide rules and log tables were the order of the day; a few mechanical calculators were in use but electronic calculators and computers were in their infancy. Many of us will remember Hewlett Packard's amazing pocket calculator with its reverse Polish notation which overnight transformed every engineer's life and enabled him to write down his answers to ten decimal places, although they were of course still based on the same assumptions.

At this time the significance of the changes in cement chemistry, which provided the rapid increase in early strength, the reduction in cement content and the consequential increase in water/cement ratio were not fully appreciated and the effects of these factors on the durability of the concrete were not understood. At a time when alkali–aggregate reactivity was unheard of in the UK, chloride attack of reinforcing steel virtually unknown, and the term rebar had not yet crossed the Atlantic, the main worries were chemical attack from sulphates, acids and other chemicals and cover to reinforcement. Determining the sulphate content of soils and groundwaters was and is relatively straightforward, and the codes provided an adequate classification system. Cover and the position of reinforcing bar were checked using the Kolectric cover meter in its handsome wooden case. This was an impressive instrument which was extremely useful for detecting steel. However, the accuracy with which cover could be measured was questionable.

1.3 ADMIXTURES

In the 1970s chemical admixtures were considered to be unnecessary products being peddled by the chemical industry to the detriment of engineering. In fact many were beneficial. However, as has been noted earlier, the only one in common use, calcium chloride, was extremely detrimental. Water-reducing agents and plasticizers were available but superplasticizers were unknown. Alternative cementitious materials such as pulverised fuel ash (PFA) and granulated ground blast furnace slag (GGBS) were not accepted for general construction and microsilica was a waste material unknown to construction. As a result the chemical analysis of hardened concrete for mix proportions was relatively simple, if rather inaccurate, since most concretes contained only Portland cement, mortar and aggregates. Today, the admixture industry has a fully operational quality management system, which we took a part in developing, and admixtures are widely used and their beneficial qualities appreciated. The cement chemist is forced to tackle concretes containing cements of different types, cement replacement materials, admixtures and additives of many forms.

Pre-war and immediate post-war cements hardened much more slowly with hydration taking place over many months, and in some cases years. The resulting concrete tended to be less permeable and therefore more durable than those produced using similar design criteria in the 1960s and 1970s. With the advent of ordinary Portland cements which were rapid hardening and gained most of their strength within the first month of life, cement contents were reduced, water/cement ratios were increased, and as a result permeability was far greater. During the 1980s and early 1990s these matters have become widely understood and mixes designed today for durability contain alternative cementitious materials which hydrate slowly over a long period of time so recreating the impermeable structure of earlier concretes whilst retaining the high strength benefits of modern cements.

1.4 REINFORCING STEEL

Whilst high tensile steel produced both by work hardening and by composition had been available for a number of years, the majority of steel reinforcement used in 1970 was mild steel. The situation is reversed today and the majority is high tensile steel which is either used uncoated or sometimes coated with protective materials such as epoxy resin.

Non-ferrous reinforcement has taken longer to develop than many thought probable in the early 1970s; however, fibre-reinforced composites are now being used for special purposes. Throughout the 1960s and for much of the 1970s, relatively porous concretes of adequately high strength were being used for structural purposes. In addition to calicum chloride, which was used in liberal quantities prior to its being banned in 1977, de-icing salt was applied in increasing quantities to the nation's highway network. As a result, highway structures such as bridges, abutments and retaining walls were saturated with salt solution in winter thus causing the problem of steel corrosion which became apparent in the mid-1970s.

STATS was involved, in 1979, in the first full-scale trial of non-destructive testing undertaken on a motorway structure in this country when it was employed on the Roundthwaite bridge by the Department of Transport. This was the first time that the half-cell potential method had been used to investigate a highway structure although it had previously been employed in connection with the offshore industry. Early equipment, which was made in-house, consisted of a single copper–copper sulphate half-cell connected to a voltmeter. Initially this was the familiar analogue voltmeter but it was soon replaced by one with a digital display. Potential maps were drawn by hand and the results were interpreted using procedures imported from the USA and based on the behaviour of bridge decks.

The probability of conditions being conducive to corrosion could be estimated and by using resistivity measurements some idea of the rate of corrosion obtained. Whilst the equipment to undertake potential mapping and half-cell evaluation has improved beyond measure, the test procedures and the interpretation of the results remain much the same and reliable means of estimating the rate of development of corrosion, possibly using linear polarization techniques, have not yet been developed for field use.

1.5 HIGH-ALUMINA CEMENT

During the 21 years since the formation of STATS there have been a number of landmarks and high-alumina cement (HAC) was the first of these. Over-reaction, largely due to lack of understanding by civil and structural engineers and over-cautious official advice resulted in the detailed examination of most, if not all, structures containing HAC; large numbers of cores were taken, Building Research Establishment pull-off tests performed, ultrasonic pulse velocity tests undertaken to estimate strength and dust samples taken for chemical analysis to determine the degree of conversion.

In fact, very few structures were found to be in need of strengthening or replacement and, whilst HAC remains on the banned list, there may be arguments for re-examination of this.

1.6 WOOD WOOL

Another problem which first surfaced in the 1970s was that related to the use of wood wool as permanent formwork. Wood wool forms were used to construct ribbed concrete floors. They were a cheap and effective means of providing economic reinforced concrete suspended floors and were quite widely accepted. Unfortunately it was discovered that the vibration energy intended to compact the concrete around the reinforcing steel was largely absorbed by the wood wool and in some cases the steel in the ribs was wholly or partially exposed. This state of affairs was concealed by the wood wool, resulting in floors of variably reduced capacity and in certain cases reinforcement free to corrode to a critical state without warning. As a result wood wool permanent shuttering was correctly banned and today is one of the ever-growing list of proscribed materials on every property solicitor's list.

1.7 DE-ICING SALT

Altogether the most significant problem with reinforced concrete which has been identified in the last 21 years is the corrosion of reinforcement caused by the use of highway de-icing salt. From being barely identifiable as a problem in the early 1970s it has become so serious a hazard that hundreds of millions of pounds have been and are being spent in identification, repair, and in some cases reconstruction. The problem arises because the chloride ion is very much smaller than the pore size found in normal concrete; Mike Grantham put this very clearly in a letter which he wrote to the NCE in the early 1980s in which he said that it was like throwing a tennis ball down a tube-train tunnel.

Where, on motorways and other major roads, there was an abundance of chloride every winter and plenty of water, penetration of chlorides to the level of the reinforcing steel took place rapidly and with scant regard for the concrete cover.

We were among the first in the field to undertake half-cell and resistivity surveys and to analyse large numbers of samples of hardened concrete. At first, studies were generally confined to investigating the condition of normally reinforced members but more recently this has been widened to include the corrosion of prestressing tendons. This has shown the need for structures to be designed to facilitate inspection using testing techniques capable of assessing the extent and form of deterioration. Such a requirement was highlighted by a moratorium on the design of new structures pending the development of suitable techniques for inspecting tendons and anchorages.

Today engineers have far greater appreciation of how to design highway structures for durability and materials scientists of how to protect reinforcement by coating and by the use of low-permeability concretes.

1.8 REPAIR

Techniques for the repair of damaged reinforced concrete have developed greatly since the early 1970s. At that time repair methods were rudimentary and generally undertaken by unskilled and largely untrained personnel. It was unusual for repair to be preceded by investigation; indeed, if investigation had been undertaken few engineers would have been able to decide what to do or would have known how to interpret the results. In consequence repairs were often ineffective with incorrect or incomplete techniques being employed.

By contrast, today, structural surveys are usually undertaken in a rigorous manner and the results analysed to identify the cause and extent of the problem and to establish possible maintenance strategies to achieve the required service life. As a result the various options can be compared and the most appropriate and cost-effective solution selected.

Techniques for the restoration of existing structures and the design of new ones have been improved by the knowledge and understanding of the processes of deterioration of reinforced and prestressed concrete. More is now known about the transport of gases, liquids and ions through concrete and associated test methods have been and are being developed. Engineers are now beginning to understand the transportation mechanisms involved (such as permeability, diffusion, absorption, etc.) and the way in which these influence durability under various environmental conditions, and some relevant performance testing techniques are being specified. The next step forward is to harness the accumulated knowledge in a form that will allow engineers to predict and achieve the desired service life of structures.

1.8.1 Cathodic protection, chloride removal and re-alkalization

Whilst knowledge of repair, by the physical removal of contaminated and damaged concrete and corrosion products before reinstatement, has grown very considerably, new techniques of cathodic protection, chloride removal and re-alkalization, all of which harness electrochemistry, have also been developed.

Cathodic protection (CP) has been employed for a long time to protect buried and submerged steel structures. Because of widespread reinforcement corrosion in unwaterproofed bridge decks in North America, US and Canadian engineers developed, in the late 1970s, CP techniques to inhibit the further development of corrosion in existing decks and so enable long-lasting repairs to be undertaken. Many bridges in the USA and Canada were protected in this way.

In this country, where most bridge decks were waterproofed, a different problem existed and research was undertaken on a wider front considering ways of protecting complete structures. Following trials, CP has been installed to protect a number of major highway and non-highway structures in this country. These developments have required the improvement of monitoring techniques, including the development of remote-sensing technology. Cathodic protection is now accepted as an appropriate technique for extending the service life of existing concrete structures in hostile environments, or of achieving the required life of some new ones.

The other electrochemical procedures are the techniques of chloride removal and re-alkalization. Chloride removal uses an applied electrical potential to force the chloride ions to the surface of the concrete where they are literally mopped up by absorbent blankets. This technique has been shown to be effective in reducing chloride at the reinforcement to acceptable levels and experience to date shows it to be long-lasting provided steps are taken to prevent further chloride penetration. Re-alkalisation was developed to counter the long-term carbonation of concrete by introducing alkalis from external sources. The long-term performance of this technique has yet to be demonstrated.

1.8.2 Repair strategies

The construction chemical industry now provides a bewildering array of protective coatings and repair systems with published performance data. Authoritative guidelines for the specification and application of patch repairs have been published and proprietary repair materials, systems and techniques are now subject to more or less rigorous accreditation procedures.

In contrast to the 1970s, the engineer today has a variety of remedial options which can be considered, including:

- Do nothing and monitor
- Artificially control the environment
- Apply protective coatings
- Undertake patch repairs
- Undertake partial or complete replacement
- Undertake strengthening to improve local factors
- Apply cathodic protection
- Undertake chloride removal
- Undertake re-alkalization.

1.9 ALKALI–SILICA REACTIVITY

In the 1970s alkali–silica reactivity (ASR) was unknown in the UK. Although it was common in many parts of the world it was widely considered by engineers in the UK to be a problem which did not concern them. Unfortunately new cement works, particularly in the south-west, produced cements with far higher alkali contents and these reacted with some siliceous aggregates resulting in the formation of silica gel and the consequential disruption of the concrete. A few structures were seriously affected, the most prominent being the Marshmills Viaduct near Exeter, and in due course this was demolished and rebuilt. However, once the full extent and nature of the problem was understood, guidelines were developed and specifications introduced to limit the total alkali content of concrete and restrict the use of certain highly reactive aggregates.

1.10 LABORATORY TECHNIQUES

During this period of great development there have been advances in the field of instrumentation for analytical chemistry but routine laboratory testing techniques have changed little. However, one technique, the petrographical analysis of thin sections of hardened concrete has proved to be of enormous benefit.

The technique relies upon experience on the part of the geologist undertaking the examination, thin sections made to the highest standard, and the use of the best optical microscope available. The experienced petrographer can identify, in addition to the types of aggregates, their grading and distribution, the presence of microcracking caused by fire damage, frost attack and the expansion and contraction of included particles. He can identify alkali–silica reactivity and many types of chemical attack. The type and distribution of cement used and the presence, type and distribution of cement replacement materials can usually be identified. The degree of hydration of the cement paste and the water/cement ratio at the time of casting can be estimated and the entrained air content of concretes can be determined. In fact, an experienced petrographer is an essential member of an investigative team.

1.11 CONCLUSION

Whilst the turmoil following the near-catastrophic failure of prestressed concrete beams made with high-alumina cement prompted us to form both STATS and Probe, it was the economic conditions at the end of the 1980s boom which brought us together and enabled us to combine the academic approach provided by Probe with that of the practical consulting engineer. STATS today embraces consultancy and laboratory services including geotechnical consultancy and field services, contaminated site evaluation, environmental audits, health and safety consultancy, building and civil engineering technology, in addition to investigations into problems with reinforced concrete. Our central laboratory services are NAMAS accredited for a wide range of British

Standards and in-house tests and we provide on-site laboratories on major construction projects.

Unfortunately, in common with many consultancies, STATS is under considerable pressure because of the overwhelming desire from many clients, in particular those in central and local government, to obtain competitive tenders and almost always to accept the lowest price. This has resulted in some markets in the public sector being effectively closed to us because we refuse to provide poor, non-professional services and believe that in the end our policy of providing good professional services at a fair price will be successful.

REFERENCES

1. Department of Scientific and Industrial Research (1934) *Recommendations for a Code of Practice for the Use of Reinforced Concrete in Buildings*, DSIR.
2. Faber, O. (1927) Paper on the plastic yield, shrinkage and other problems of concrete and their effect on design. Paper presented to the Institution of Civil Engineers on November 15.
3. British Standards Institution (1957) Code of Practice for Reinforced Concrete, CP114, BSI, London.
4. Glanville, W.H. and Scott, W.L. (1973) *The Explanatory Handbook on the BS Code of Practice for Reinforced Concrete CP114, 1957* (incorporating 1965 amendment to the code).
5. British Standards Institution (1965) Amendment to Code of Practice for Reinforced Concrete, CP114, BSI, London.
6. British Standards Institution (1972) *Code of Practice for the Structural Use of Concrete CP110*, BSI, London.
7. Neville, A. (1972) *Properties of Concrete*.
8. Concrete Society (1990) Report of working party on changes in the properties of ordinary Portland cement.

Discussion of Chapter 1

Dr Adam Neville John entirely justifiably accused me of being party to the school of thought that there is nothing wrong with chloride, within certain limitations, being put into concrete. I mean, it is interesting to look at such an erroneous attitude being so widespread, but what I wanted to say is that in [the] early 1970s many people, including myself, became aware of the fact that chloride put into concrete led to corrosion. At that time I was on the Codes Committee, the committee which wrote CP110 which was the new code for design of reinforced and prestressed concrete using the limit state approach and I raised the question of allowing calcium chloride; I had support from the chairman, the late Dennis Matthews, and I thought we were moving somewhere but then we had representation from ICI who I believe were then the only manufacturer of calcium chloride. That was the time when everybody was terribly concerned about imports and under the circumstances we agreed to do nothing. Well, as we were told by John (Glanville), in 1977 we, the same committee, passed an amendment banning calcium chloride. I can't say we have lived happily ever after because in many structures chlorides ingress from outside, but that is another topic.

Now has anybody been provoked into saying more, on this or any other topic?

Mike Grantham, *MG Associates* I was the one who penned the comment on chlorides referred to by John Glanville. The thing that concerns me is that we still have no code of practice on this subject. However, highway structures engineers now generally understand the transport mechanism for de-icing salts. Several pints of salt contaminated water are dropped on car parks by every vehicle coming in during winter months. Our code of practice suggests that 50 mm of cover in a 40-grade concrete is sufficient proof against that but several authors recently have published work on this. Phil Bamforth of Taywood Engineering wrote a very good article in *Concrete* on 'Admitting that chlorides are admitted', which suggests that design to the BS 8110 'very severe design' category will give a life of about nine years before problems with chlorides commence. I have been doing some work on this subject on a legal case at the moment which shows that in a car park built about seven years ago the problems began at about five years.

Dr Adam Neville Does anyone want to comment on codes of practice?

Mike Walker, *Concrete Society* Coming to the question of the dangers of using chlorides, these have been debated for a very long time. The use of chlorides was in and out of favour for some time before the seventies. In the CIRIA book on maintenance and repairs there is a paper by Darrell Leek, now Mott Macdonalds, and myself which plots the history of attitudes toward chlorides both in the UK and in other parts of the world. In this review the way in which attitudes to the use of chlorides fluctuated up until the mid to late seventies can be clearly seen. With regard to the point that John (Glanville) made about calcium–alumina cement, there is a Concrete Society working party which is examining the place of calcium–alumina cement in the industry. By no means is the working party likely to suggest that it can be used generally but it intends to determine where its beneficial qualities could be used in specific situations without any problems.

Dr John Newman I could not let this occasion pass without mentioning something more about a subject in our paper, namely, calcium chloride. The origin of the 1.5% calcium chloride limit is in the 1940s and centres on a BRE paper concerned with the relationship between the corrosion of steel reinforcement and the amount of calcium chloride in the concrete surrounding the steel. In those days, and I hope it is different now, limits appeared to be drawn separating black (unacceptable) from white (acceptable). In the case of calcium chloride the limit was 1.5% of anhydrous material and above this value lay disaster and below it satisfaction. The limit was determined by testing for approximately one year which was considered a long period of time to wait for results. I have to say that the author of the paper was a certain A.J. Newman, my father, and the limit remained in force until the 1970s. I often wonder how many people who read that paper appreciated the basis on which the limit had been derived and questioned the lack of a grey area. As we know now there is a very large area of grey and one year is a very short time in comparison with the anticipated service lives of structures.

Dr Adam Neville We shall have time for more discussion later. At the moment I would like to move to the next paper, which is written by Professor Bungey.

2

Non-destructive testing: assessment of concrete durability related to expected performance in service

by J.H. Bungey

ABSTRACT

The current situation concerning non-destructive testing techniques is reviewed in the light of recent developments in apparatus, procedures and interpretation. This includes measurement of cover to reinforcing steel; corrosion risk and corrosion rate assessment for reinforcing steel; location of internal voids, disruption and deterioration; location and assessment of internal moisture and chloride contamination; and abrasion resistance and surface zone strength assessment. Techniques considered encompass cover meters, electrical corrosion assessment methods, ultrasonics, thermography, radar, radiography, dynamic response, relative humidity meters and partially destructive strength tests. Reference is made to appropriate guidance documentation, including standards and recently published research, and the material is set in the context of practical durability-related concerns of the construction industry.

Keywords: Non-destructive, testing, concrete, durability, *in situ*.

2.1 INTRODUCTION

Non-destructive testing of concrete has seen widespread developments over the past 25 years. Twenty five years ago the range of available techniques in general use was largely restricted to rebound hammer, pulse velocity, radiography or initial surface absorption methods. The need to assess high-alumina cement (HAC) concrete in the 1970s highlighted the need for a wider ranging armoury of tests related to *in situ* strength assessment. Apart from the internal fracture tests initially aimed specifically at HAC, a range of partially destructive techniques has now evolved which permit estimates of near-to-surface *in situ* strength. Alkali–aggregate reactions and reinforcement corrosion have been two of the principal causes for concern in more recent years together with uncertainties about grouting deficiencies and tendon corrosion in post-tensioned construction. It is not surprising therefore that the principal emphasis of *in situ* testing has swung towards test methods capable of assessing parameters associated with durability and integrity. This has led to developments and refinements in testing procedures, apparatus and interpretation. Usage still, however, tends to be dominated by the trouble-shooting role whilst increased application during construction grows only slowly. The range of currently available tests is very wide as illustrated by Table 2.1. Details of methods and procedures are described in detail elsewhere, including the third edition of the author's book [1].

Particular attention should be given to the benefits of combinations of test methods rather than their use in isolation. Some methods can provide valuable preliminary information prior to the use of other techniques, whilst in different situations a significant increase of confidence in both interpretation and prediction capabilities may be possible by comparing and combining results.

Documentation, in the form of standards and authoritative technical reports has developed steadily, with a major revision and extension of British Standards to form the BS 1881 pt 200 series during the 1980s and early 1990s. Some major gaps have been plugged by the Concrete Society, who are currently working on further reports dealing with reinforcement corrosion assessment and subsurface radar.

2.2 EQUIPMENT DEVELOPMENTS

There has been a general trend towards digitization, automation and electronic data storage. This, to varying degrees, encompasses established methods such as rebound hammer testing and cover measurements, as well as surface-zone permeability and reinforcement corrosion tests. Whilst it is a fashionable development which may undoubtedly make some aspects of work on site easier, especially in situations with difficult access, it is important to recognize that this must be balanced against increased cost and a temptation to believe that results are more accurate. In some cases a loss of control and 'feel' by the operative may be an important negative factor.

In some areas of testing, such as cover measurement, there is now a

Table 2.1 Basic characteristics of principal test methods

Property under investigation	*Test*	*Equipment type*
Corrosion of embedded steel	Half-cell potential	electrochemical
	Resistivity	electrical
	Linear polarization resistance	electrochemical
	Cover depth	electromagnetic
	Carbonation depth	chemical and microscopic
	Chloride penetration	chemical and microscopic
Concrete quality, durability and deterioration	Surface hardness	mechanical
	Ultrasonic pulse velocity	electromechanical
	Radiography	radioactive
	Radiometry	radioactive
	Permeability	hydraulic
	Absorption	hydraulic
	Moisture	chemical and electronic
	Petrographic	microscopic
	Sulphate content	chemical
	Expansion	mechanical
	Air content	microscopic
	Cement type and content	chemical and microscopic
	Abrasion resistance	mechanical
Concrete strength	Cores	mechanical
	Pull-out	mechanical
	Pull-off	mechanical
	Break-off	mechanical
	Internal-fracture	mechanical
	Penetration resistance	mechanical
	Maturity	chemical and electrical
	Temperature match curing	electrical
Integrity and structural performance	Tapping	mechanical
	Pulse-echo	mechanical/electronic
	Dynamic response	mechanical/electronic
	Thermography	infra-red
	Radar	electromagnetic
	Reinforcement location	electromagnetic
	Strain or crack measurement	optical/mechanical/electrical
	Load test	mechanical/electronic/electrical

large choice of apparatus which is commercially available whilst in others the range is far more restricted. There have been recent advances in the development of equipment for thermography, subsurface radar and impact echo testing which have greatly enhanced operational capabilities, but which in some cases have not yet been fully developed. The wide range of equipment available in the UK was reviewed in 1992 in Technical Note 143 of the Construction Industry Research and Information Association (CIRIA) [2] and is constantly increasing.

2.3 TEST METHOD DEVELOPMENTS

2.3.1 Strength testing

The pull-out and pull-off techniques have proved to have the greatest potential although their application within the UK is still very limited. Standardization of procedures by BS 1881 pt 207 [3] in 1992 has been an important step forward for these methods which measure strength-related properties of the all-important cover-zone concrete. Whilst specific strength correlations are recommended for the concrete under test, these are sensitive to far fewer variables than rebound hammer and pulse-velocity testing, which have a predominantly comparative role.

Strength predictions from partially destructive methods are unlikely to be as reliable as those from cores but results are available instantly, and damage and time requirements are reduced, hence increasing possibilities for wider coverage. Use of simple tests to find the most effective location for more damaging techniques such as coring may be particularly worthwhile.

Cast-in pull-out testing (Fig. 2.1) is the most reliable approach for direct *in situ* testing at very early ages and low strength. Figure 2.2 shows results obtained by the author with reference to cooling tower construction at strengths as low as 2 N/mm^2 which illustrates the excellent correlations that are possible. This method has considerable potential for use during construction, having been used successfully on the Great Baelt project in Denmark. It can also be used successfully up to about 130 N/mm^2 cube compressive strength.

Whilst use of direct maturity measurements to monitor *in situ* strength development has received little acceptance in the UK, interest in North America has been greater [4] and developments of understanding and interpretation are still proceeding. Combination with pull-out tests may be particularly useful [1]. Temperature-matched curing has attracted greater attention in the UK and the current BS DD92 [5] is being revised and upgraded to become BS 1881 pt 130.

For *in situ* strength assessment of existing structures it is surprising that the 'Capo' version of the pull-out test (Fig. 2.3) has not become

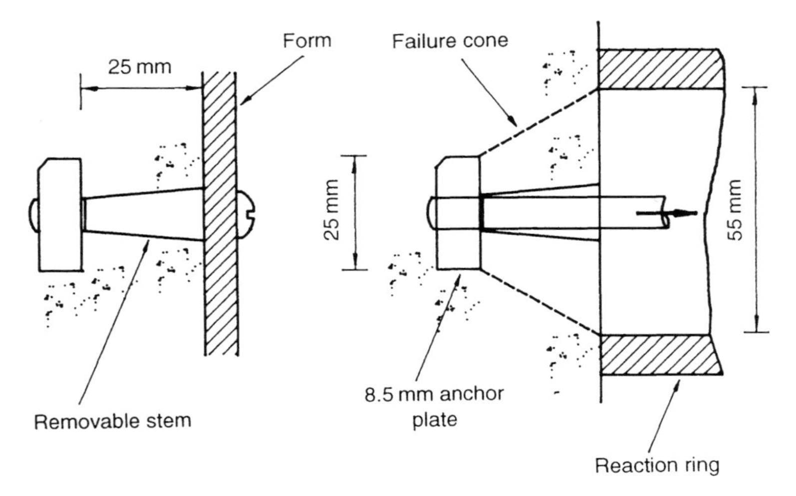

Figure 2.1 Cast-in pull-out test.

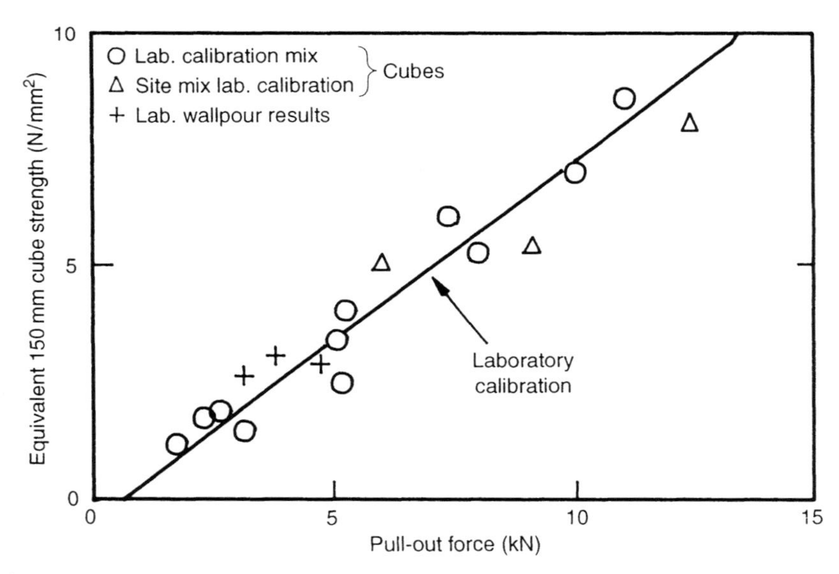

Figure 2.2 Low strength pull-out results.

more widely used despite successful application to high-strength concrete on the Channel Tunnel project. Although Windsor probe and internal fracture methods have their place, pull-off testing has probably received greatest recent attention. Bond testing of repairs to concrete is an application for which this method is particularly suitable [6] when partial coring is used, as illustrated in Fig. 2.4.

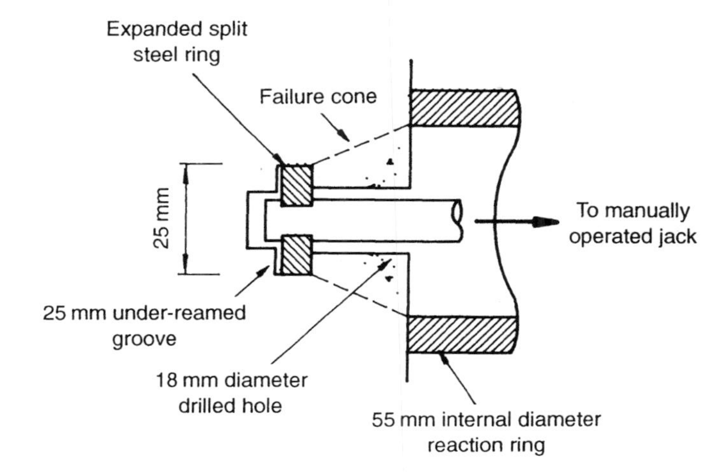

Figure 2.3 Capo test configuration.

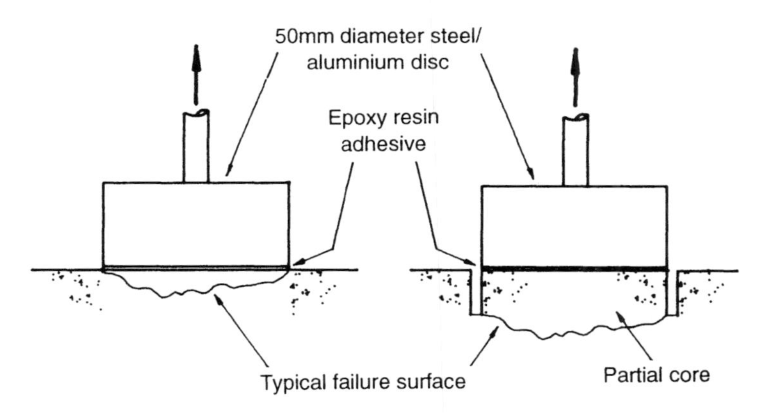

Figure 2.4 Pull-off test.

Standardization of the test configuration has been especially important for this method since a wide range of versions exists worldwide [7] and the author [8] has demonstrated the extent to which a range of factors, including disc characteristics, drilling depth and load rate, may influence results. Figure 2.5 illustrates, for example, the effects of disc proportions and materials on surface tests. Experimental and finite-element studies have shown that where partial coring is used, whether on plain concrete or for testing repairs, a drilling depth of at least 20 mm into the base material is needed. The effects of disc proportions are however less

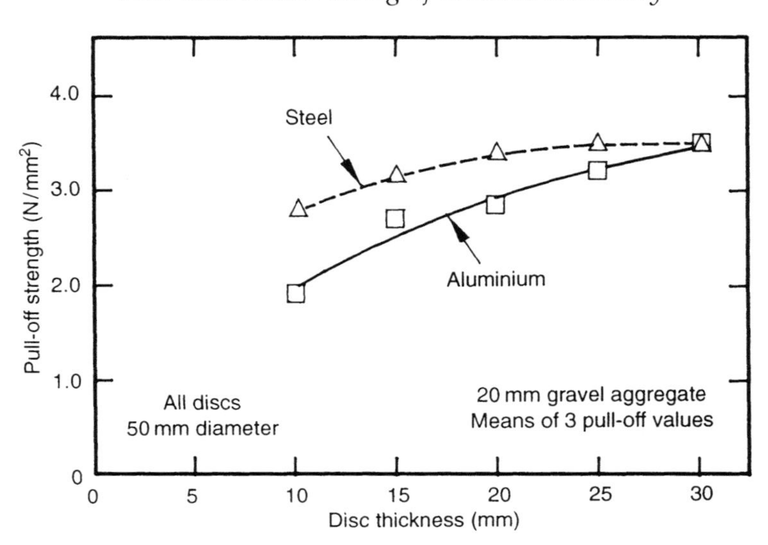

Figure 2.5 Effects of disc properties on surface pull-off tests on concrete. (Reproduced from Bungey, J.H. and Madandoust, R., *Magazine of Concrete Research*, 44 (158), 21–30; 1992.)

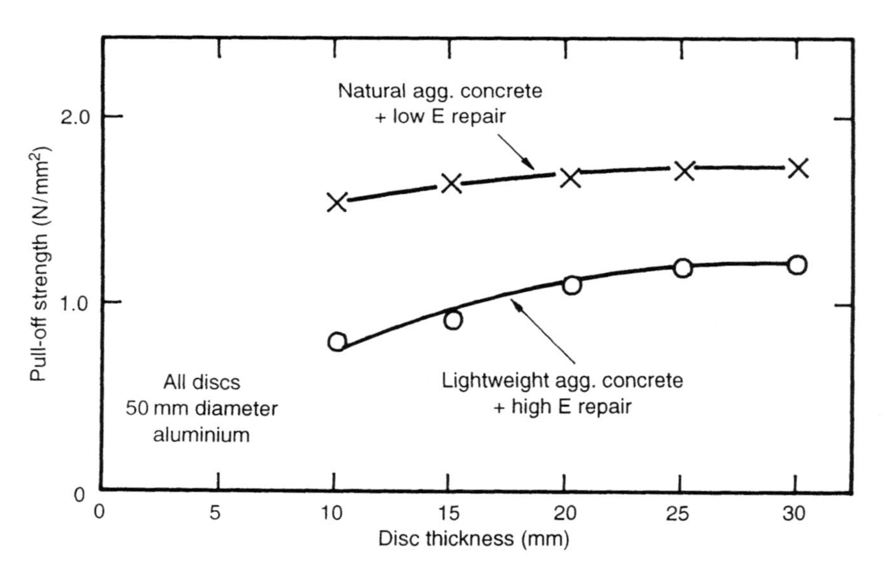

Figure 2.6 Effects of disc thickness on partial-cored pull-off tests on repairs.

significant where partially cored tests are used on repairs, as illustrated in Fig. 2.6 for two extreme combinations of substrate and repair elastic properties.

Other aspects of *in situ* strength assessment which are worthy of note include recent work on the application of *in situ* tests to lightweight concretes as well as a growing interest in *in situ* strength assessment of high-strength concretes. In the case of lightweight concretes it has been demonstrated that most test methods can be used successfully but that strength correlations change with lightweight aggregate type and the presence or otherwise of natural fine aggregates as illustrated by Fig. 2.7.

Variability of results is generally less than for normal concretes. The effect of cement replacements on traditional procedures for interpretation of core results is currently being addressed by the Concrete Society in a revision of their well-established report on core testing, *Concrete Core Testing for Strength* [9a].

The use of statistical procedures to assist interpretation of reliability of *in situ* strength estimates from small samples, as well as the determination of the optimum number of tests to undertake, has attracted much attention in the USA and Denmark. United States procedures are covered by the report by ACI Committee 228 [10] which is currently being revised, whilst Leshchinsky [11] has recently reviewed worldwide procedures. There is a difficulty in transferring US procedures directly to

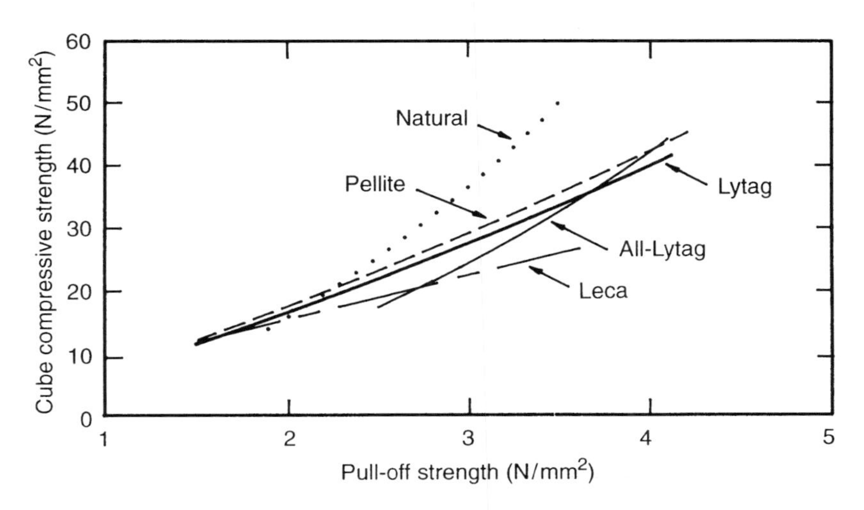

Figure 2.7 Typical pull-off strength correlations for a range of aggregate types. (Reproduced from Bungey, J.H. and Madandoust, R., *Proceedings of the Institution of Civil Engineers: Structures and Buildings*, **104**, 275–83; 1994.)

UK practice because of differences in commonly adopted confidence limits, but CIRIA Technical Note 143 [2] provides some guidance related to current UK practice. The lack of worldwide agreement on acceptance criteria is possibly one of the major impediments to the increased use of *in situ* strength testing for compliance purposes during construction, and RILEM Committee 126 IPT is currently addressing this problem.

2.3.2 Permeability testing

Permeability is widely regarded as a key parameter influencing durability performance. A substantial number of approaches have been examined at the research level aimed at measuring some permeability-related properties of surface-zone concrete, although differences between the surface and the interior due to construction operations such as vibration, finishing and curing must be recognized. These methods include sorptivity and surface water absorption, as well as air and water permeability. Results of such tests may be combined with those described in section 2.3.3 to assist performance predictions relating to reinforcement corrosion as indicated below.

The only standardized test in the UK which is at present suitable for *in situ* usage is the long-established initial surface absorption test which is a refinement of the simple qualitative water-drop approach. This involves clamping a watertight cap to the undisturbed concrete surface and measuring the flow rate into the surface over a period of up to two hours under a 200 mm head as shown in Fig. 2.8. Despite some

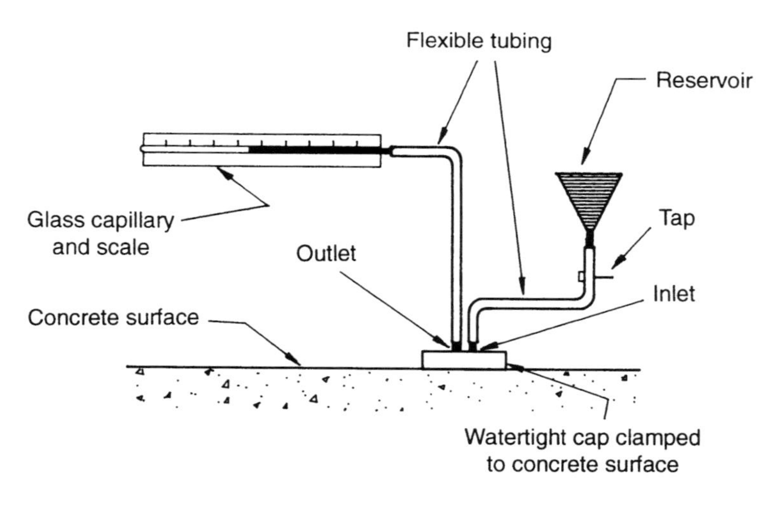

Figure 2.8 Initial surface absorption test.

potential practical difficulties of field application, this low-pressure test has continued to be used over many years, although the revision of BS1881 pt 5 [12] as BS 1881 pt 208 has suffered long delays because of the absence of suitable precision data. This proposed revision includes allowances for temperature effects and it is to be hoped that the matter will be resolved before too long.

The CLAM and AUTOCLAM systems offer pressurized versions [13] with the capability for an air test as well as a water test, whilst developments of an alternative method developed some time ago at the Building Research Establishment by Figg have led to the 'Poroscope' air and water permeability apparatus. This latter technique involves a small hole drilled into the surface to create a sealed cavity within which either a vacuum or water pressure regime is created.

Other techniques that have been proposed [1] are generally based on some variation of the above principles and include attempts to overcome the dominant problem of the influence of *in situ* concrete moisture conditions. These include developments of the guard-ring principle and vacuum-drying techniques to standardize test conditions, as well as other conditioning methods. Sensitivity to moisture varies between tests and it is claimed that water tests at higher pressure, such as with the CLAM systems, are less sensitive to variations than air tests or water at lower pressures. Measurement of *in situ* internal moisture conditions may assist interpretation of permeability and other results but is difficult and generally requires drilling holes. The most commonly used methods are indirect, being based on measurements of the relative humidity within a sealed hole and involve chemically based meters, electrical capacitance and conductance measurements, or dew-point measurements [2]. The latter approach uses an electronically controlled cyclic chilled-mirror principle and seems to offer advantages of speed and reliability compared with other methods. The condensation is detected by its effect upon a light beam reflected from the mirror and the relative humidity evaluated from a knowledge of the temperature involved. Thermography and radar which are described below can be used to detect moisture comparatively, but microwave absorption offers a potentially direct quantitative approach for the future.

In Europe the matter of permeability testing related to durability has recently been addressed by RILEM Committee 116 PCD, which is expected to report shortly [14], and an international project is under way supported by the European Union Measurements and Testing Programme to provide data for a range of test methods.

Various proposals have been made to link different permeability-related measurements to durability performance [13, 15] but the categories are very broad and unlikely to be refined until the testing procedures are more clearly defined. Even then, interpretation will still need to encompass the particular characteristics of the wide range of modern concrete compositions.

2.3.3 Reinforcement corrosion testing

Estimation of the thickness of cover concrete is an important factor in predicting the likelihood of future corrosion, and several models have been proposed for combining such values with chloride diffusion and carbonation characteristics. Those linking carbonation-induced corrosion with cover and permeability [16] are further advanced than those involving chlorides. Various cover meters are available [2] based on either magnetic induction or eddy current principles and with wide-ranging operational capabilities. The factors affecting the accuracy of cover meter results are widely recognized [17] but the need for realism concerning accuracies of depth prediction under site conditions is less widely accepted. In particular, the importance of on-site calibration cannot be overemphasized. The confidence levels of such measurements are obviously critical in determining the reliability of lifetime predictions into which they are incorporated.

Within the last few years subsurface radar [18] has emerged as a useful alternative technique for location of steel reinforcing bars and ducts and has a potential depth range well in excess of most cover meters. This method also has its limitations, however, and depth estimation again requires on-site calibration or detailed knowledge of materials' properties. Other techniques involving magnetic and ultrasonic imaging are, as yet, at the research laboratory stage.

The use of half-cell potential methods to assess corrosion risk by electrochemical means is well established [19] and widely used (Fig. 2.9). Although no British Standard is available, there is a long established ASTM Standard [20] and the Concrete Society is producing a technical report in collaboration with the Institute of Corrosion addressing this technique. Classification is in broad bands of corrosion risk and the method is good at locating regions of different risk on a comparative basis. This approach addresses the risk of corrosion activity, and if supplemented by measurements of concrete resistivity [21] can be used to assess the potential severity of corrosion in high-risk areas. Resistivity measurement may be based on a two-probe or non-damaging four-probe technique shown in Fig. 2.10. Equipment is commercially available for both these tests and research has addressed the influence of physical and environmental factors upon interpretation in the field. Half-cell potential measurements have been shown to be affected by the moisture state of the surface and by the presence of chloride contamination, whilst temperature has been found to be particularly important for resistivity values as illustrated in Fig. 2.11. Other factors to be

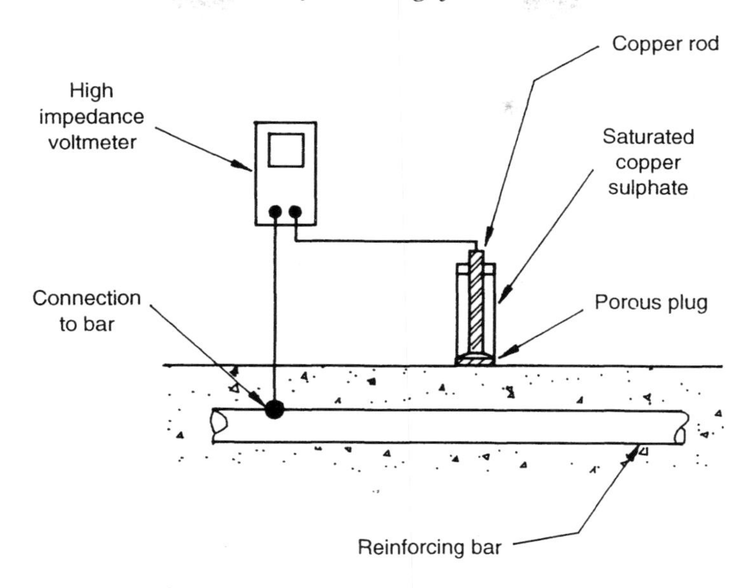

Figure 2.9 Half-cell potential test. (Reproduced from Vassie, P.R. (1991) *Application Guide 9*; TRRL.)

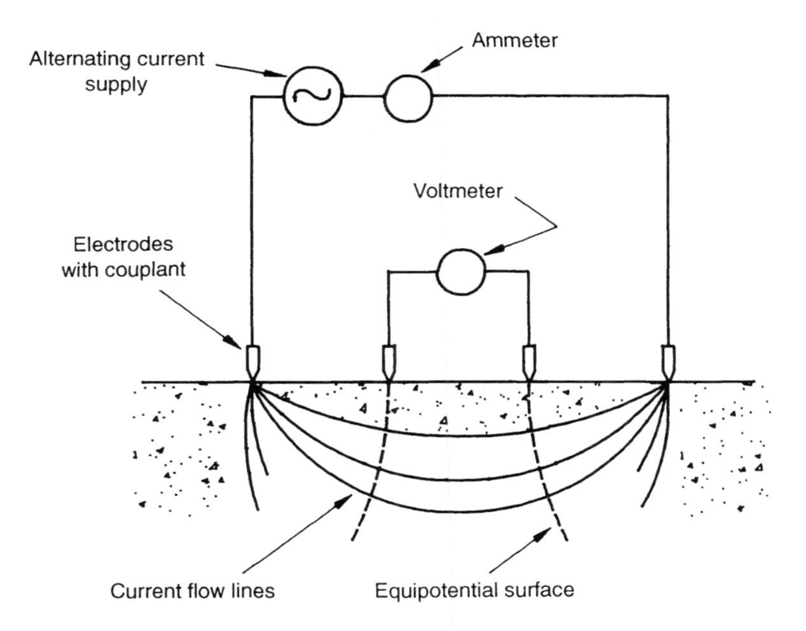

Figure 2.10 Four-probe resistivity measurement.

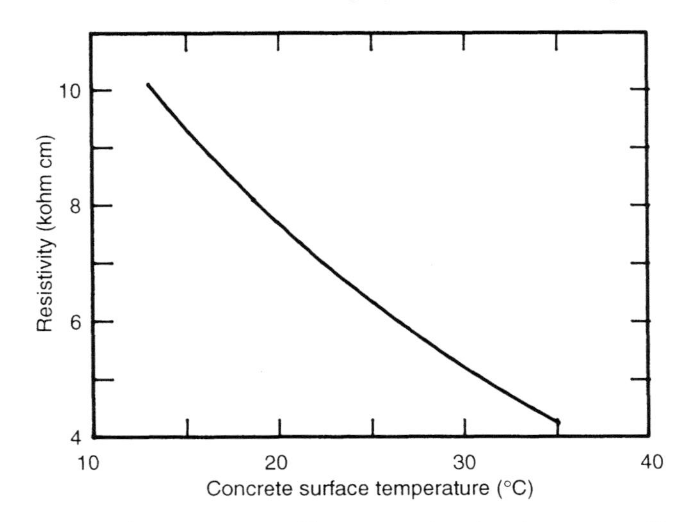

Figure 2.11 Typical temperature effects on resistivity of concrete. (Reproduced from Gowers, K.R., Millard, S.G. and Bungey, J.H. *Proceedings of International Conference on Non-destructive Testing in Civil Engineering*, vol. 2, pp. 633–657.)

considered include the presence of steel reinforcement close to the measurement zone, surface layers due to carbonation or wetting, element size and edge distances.

The principal limitations of these methods are that they cannot indicate the current corrosion state of reinforcing steel or the rate at which it may be occurring. A variety of perturbative techniques have been examined over the past few years to address this issue. These include:

- AC impedance
- AC harmonic analysis
- linear polarization resistance
- galvanostatic pulse.

Of these, linear polarization resistance appears to be the most promising [23] and commercially available equipment has received extensive trials in the USA as part of the Strategic Highways Research Programme. The technique illustrated in Fig. 2.12 involves the application of a small electrochemical perturbation to the steel via an auxiliary electrode placed on the concrete surface.

This method estimates current corrosion rate but cannot predict past or future rates. One particular difficulty is establishing the area of steel which is influencing the test, and a guard-ring approach can be used to attempt to overcome this problem, whilst a further uncertainty concerns the uniformity of corrosion around the bar perimeter.

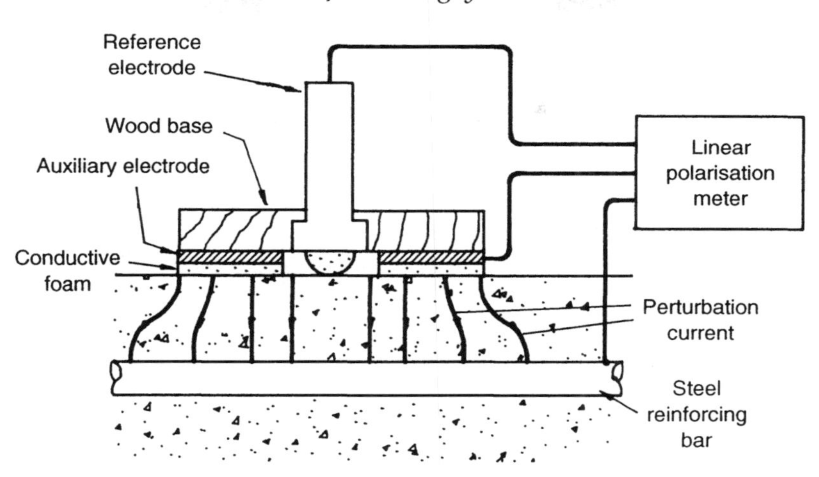

Figure 2.12 Linear polarization resistance test.

Linear polarization apparatus is available in the UK but field experience is, as yet, relatively limited. It is to be hoped that future research will lead to improvements in the ability to assess corrosion conditions non-destructively and such work is under way at a number of centres.

Prestressing steel in post-tensioned construction presents even more severe assessment problems since the presence of a metallic duct prevents the use of these electrochemical techniques. Whilst research is looking at the possibilities of electrical or ultrasonic techniques involving propagation along the length of the strands, current field methods are predominantly based on invasive visual examination.

2.3.4 Other durability related testing

Measurement of abrasion resistance is an area which does not appear to have attracted much attention since the work of Sadegzadeh and Kettle [24] although they have indicated the potential value of non-destructive methods including rebound hammer and initial surface absorption tests in this respect. Although there are several standardized abrasion tests around the world, few are suitable for use *in situ* and none has gained a foothold in the UK. Chaplin [25], however, has provided some classification guidelines for industrial floors based on an accelerated wear test. One area that has seen recent development is the use of optical fibres for structural monitoring purposes [26]. This is particularly suitable for detecting movement over considerable lengths of an element or structure.

There are no suitable direct *in situ* tests for freeze–thaw resistance, but chemical and petrographical methods applied to samples removed from the concrete have a valuable role to play in determining this property, as well as carbonation depths, chloride profiles, sulphate contents, alkali–silica reactivity and other key materials' characteristics for use in durability performance predictive models. Diffusion and similar tests on samples removed by coring can also yield important data for predictive models. Developments are hampered by lack of agreement concerning appropriate allowances for chloride binding and maturity effects for a particular model and diffusion coefficient.

2.3.5 Integrity and performance testing

This is an area that has seen many new developments aimed at improving the ability to detect buried features which may influence future durability. None of these techniques offers a total guarantee of success and many reports emphasize the value of corroborative evidence provided by appropriate combinations of methods. Reflective-mode ultrasonics has been developed to the point where it can be used for quality checking of precast concrete components during manufacture but this is not yet at a stage at which it can be applied to *in situ* situations. Thickness gauges have been developed but are not commercially available in the UK.

Dynamic response tests have grown in popularity and range from localized surface impacts to large-scale vibrations of elements or structures.

Modal analysis, transient response and spectral analysis of surface waves (SASW) are all techniques in which responses at some position away from the impact point are measured, processed and analysed to compare defects and layering effects [27]. Whilst usage is growing, most reported *in situ* applications are still largely developmental. One technique that has attracted considerable attention in the USA and for which portable apparatus is commercially available [2] is based on frequency-domain analyses of the reflected signals resulting from a surface impact. Features such as voids or delamination can be detected by a shift in the amplitude of higher-frequency components of the return signal. Adjustment of the duration of the input impulse can be used, for example, to enhance or supress the response from a layer of reinforcing bars [28]. This 'impact–echo' technique has been used to monitor the thickness and integrity of concrete slabs and walls as well as cracking in beams, and studies to detect air voids resulting from inadequate grouting of steel post-tensioning ducts are promising [29]. The use of neural networks to assist interpretation has also been considered [30].

Infra-red thermography has been in use for many years to detect hidden surface-zone cracking or moisture by comparative study of surface-temperature differentials. Recent improvements in field equipment have greatly increased the sensitivity with which measurements can be made, leading both to greatly improved sensing and far greater flexibility in operating conditions. This has been achieved by using closed-cycle electric micro-coolers, which work down to temperatures as low as −196°C in place of liquid-nitrogen cooled detectors, to provide a long-wavelength system which can be operated safely under a wide variety of site conditions. This provides for more accurate results, enabling lower surface temperature differentials to be identified with reduced solar interference, so enhancing operating capabilities and enabling daytime sensing to be undertaken under certain conditions. Battery-operated equipment is portable (Fig. 2.13) and can give quantitative temperature data on the displayed image. The image is recorded on computer floppy disk which can subsequently be processed and analysed in the office. The operation of computer processing and analysis enables far more information to be gained than was possible

Figure 2.13 Infra-red thermography apparatus. (Reproduced with permission from Rollinson-Glanville Consultants.)

with earlier liquid-nitrogen cooled equipment which relied upon recording for subsequent video play back. The technique is particularly useful for remote-sensing situations such as the detection of cladding defects, delamination of renders and applied façade treatments and has been used widely, particularly in the USA, for investigation of delamination in reinforced concrete highway pavements.

Subsurface radar is a technique that has emerged over recent years as a useful tool for testing structural concrete and is still growing in popularity. It is non-damaging and only requires access to one surface, thus enabling large areas to be inspected relatively quickly. The general principle is illustrated in Fig. 2.14 and successful applications include determination of construction features and element thickness; location of buried features such as voids, cracking, pipes and reinforcing steel; and location of moisture and salt contamination [31]. At present these are largely comparative and require some calibration drilling or breaking out. In North America there has been much interest in high-speed vehicle-mounted operation for highway pavement and bridge deck

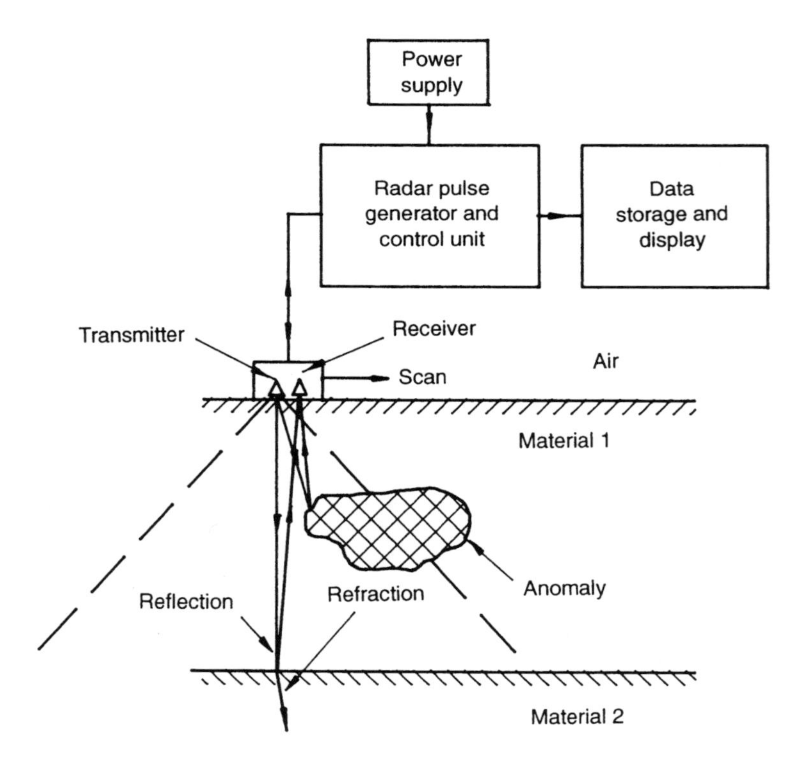

Figure 2.14 Subsurface radar.

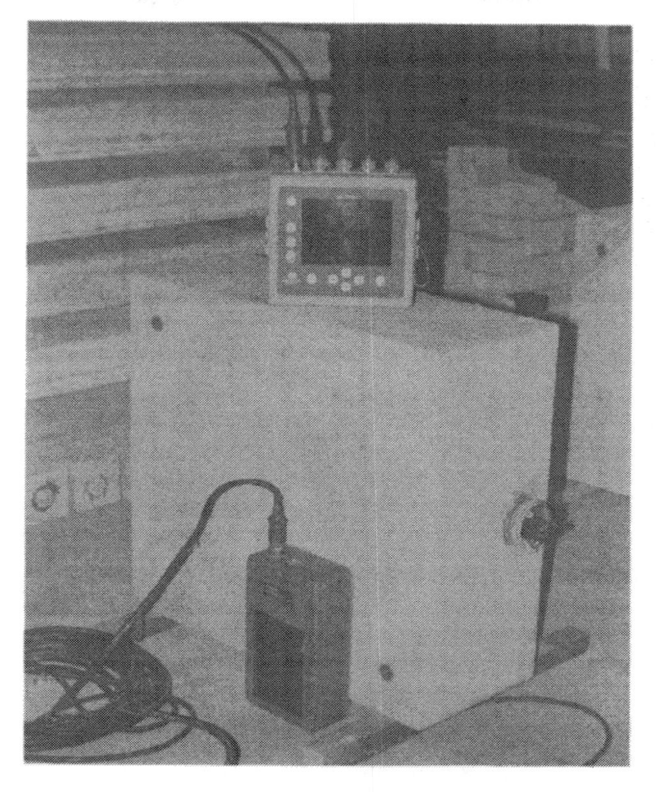

Figure 2.15 Radar equipment.

integrity assessment and this facility is now also available in the UK. Equipment developments have led to readily available, portable digital apparatus with colour display and signal processing capabilities (Fig. 2.15).

A series of waveform pulses are transmitted into the concrete surface and results are generally presented in the form of a plot of the reflected signals obtained during a scan, with amplitude and polarity identified by varying grey-scales or colours. A typical result obtained during a scan over a rectangular air void is shown in Fig. 2.16. Contrast of materials' properties means that water-filled voids or cracks are more readily detectable than air-filled ones.

Interpretation relies on characteristic pattern recognition and, although signal processing can clarify the images, there is a limit to what can be achieved with results from inappropriate apparatus settings. Where multiple features are involved, such as closely spaced reinforcing bars, it

Figure 2.16 Radar scan over rectangular void in concrete.

becomes very difficult to identify particular features of interest. Interpretation can, however, be assisted by the study and comparison of individual reflected waveforms. It is possible to estimate the depth of features if the relative permittivity of the concrete is known, since this affects signal velocity. Signal attenuation can similarly be related to the electrical conductivity of the concrete. Quantitative interpretation relies upon a knowledge of these two properties which are both heavily influenced by the moisture content. Conductivity is further influenced by the presence of salts such as chlorides.

A major research project by the author has recently attempted to quantify these properties for a range of concretes and moisture conditions using typical signal frequencies within the 1 MHz to 1 GHz range. Detailed results of this work will be published shortly.

One application of radar that has proved to be useful is the location of metallic prestressing ducts within beams. Unfortunately the steel ducts mask the interior, thus voiding in the grout cannot be found, but research results suggest that, at least in the laboratory, major voids in plastic ducts may be detectable. Field investigations are, however, likely to be hampered by the presence of reinforcing steel around the ducts. This particular problem has led to the development of one of the few new field test methods in the form of vacuum–pressure testing to estimate the magnitude of voids, and this approach is growing in use.

Concern over voids in grouted metal ducts in existing structures, and the consequent effects upon durability, has also led to a recent resurgence of interest in the use of radiography. This method is fundamentally slow and requires extensive safety precautions, but a French Scorpion lorry-mounted system has been successfully used in the UK [32]. A real-time video image is obtained as the source and detector are tracked along the length on opposite faces of a beam web. Use is obviously limited by practical constraints and an extensive exclusion zone is necessary from the health and safety point of view. An alternative approach using a 6 MeV betatron system has also been described which offers a reduced level of risk and it is claimed that voids in the order of 15 mm × 15 mm are detectable [33].

2.4 DOCUMENTATION DEVELOPMENTS

Many documentation developments, including those by British Standards, the Concrete Society and CIRIA, have been identified previously. In addition to these, RILEM has recently produced a draft document relating to damage classification [34] which could prove a useful step towards developing standardized categorization of the condition of concrete structures for durability and performance predictions. Whilst some draft ISO standards have been prepared, international standardization progress is generally slow, especially in Europe.

There have been some reports of efforts to develop expert systems to assist non-destructive testing, but a view seems to be emerging that their role is likely to be confined to that of a training aid to supplement written documentation. This is because of the complexity of interpretation of non-destructive testing combined with the level of programming investment required.

2.5 CONCLUSIONS

Despite a large amount of research activity over recent years, there are few genuinely new non-destructive test methods available to assist predictions of durabilty and lifetime performance. Most attention has been paid to the development of procedures for, or modification of, methods of testing and interpretation that have been available in some form for many years to provide information on specific parameters of the concrete. Much of the effort has been duplicated, often unintentionally, due to the overwhelming volume of literature available which is often overlooked by new researchers. Lack of standardization of procedures in some areas, such as permeability assessment, has hampered acceptance and development of interpretation beyond the specific measured parameter.

Most equipment developments have involved computerization in some form leading to increased cost without any particular increase in usefulness of the results obtained. The principal exceptions to this are dynamic response, thermography and radar testing, which all yield comparative integrity-related information.

Non-destructive testing can, however, if properly planned and executed, yield much information to enable predictions of future performance to be made. The importance of combinations of test methods rather than their use in isolation must be emphasized. This may encompass the use of simple comparative methods prior to those which may be more costly or damaging (but yield more reliable data) as in strength testing, or the use of complementary methods to enhance confidence in findings and predictions. Some examples have been given above to illustrate how one test method may provide only one part of the jigsaw whilst others may enable it to be completed more fully. This is particularly true where durability and integrity are concerned.

It seems unlikely that there will be any major breakthroughs in terms of increased accuracy of most areas of testing in the near future or even in the longer term. This is despite improvements in the development of allowances for environmental effects and the possible use of neural networks. Attention must instead be concentrated on the more difficult problem of developing appropriate ways of making best use of those results that can be obtained to predict future performance in service. This will always be a major challenge due to the diverse and inherently variable nature of concrete and the complex inter-relationships between performance and the environment. Extensive research to address this problem is certainly still needed.

ACKNOWLEDGEMENTS

Thanks are due to many colleagues, including Dr S.G. Millard at the University of Liverpool, for their co-operation over many years. The financial support of the former Science and Engineering Research Council for work on some aspects of the subject is also gratefully acknowledged.

REFERENCES

1. Bungey, J.H. and Millard, S.G. (1995) *Testing of Concrete in Structures*, 3rd edn, Blackie, Glasgow.
2. Bungey, J.H. (1992) *Testing Concrete in Structures – a Guide to Equipment for Testing Concrete in Structures*. Tech. Note 143, Construction Industry Research and Information Association, London.
3. British Standards Institution (1992) *Testing Concrete: Recommendations for the Assessment of Concrete Strength by Near-to-surface Tests*. BS 1881 pt 207, BSI, London.
4. Carino, N.J. (1991) The maturity method. *Handbook on Non-destructive Testing of Concrete* (eds. V.M. Malhotra and N.J. Carino), CRC Press, Boston, pp. 101–46.
5. British Standards Institution (1984) *Method of Temperature Matched Curing of Concrete Specimens*, DD92, BSI, London.
6. Cleland, D.J. (1993) In-situ methods for assessing the quality of concrete repairs. *Proceedings of the Institution of Civil Engineers: Structures and Buildings*, **99**, 68–70.
7. McLeish, A. (1992) *Standard Tests for Repair Materials and Coatings – Part 1: Pull-off Testing*. Tech. Note 139, Construction Industry Research and Information Association, London.
8. Bungey, J.H. and Madandoust, R. (1992) Factors influencing pull-off tests on concrete. *Magazine of Concrete Research*, **44** (158), 21–30.
9. Bungey, J.H. and Madandoust, R. (1994) Evaluation of non-destructive strength testing of lightweight concrete. *Proceedings of the Institution of Civil Engineers: Structures and Buildings*, **104**, 275–83.

9a. Concrete Society (1987) *Concrete Core Testing for Strength*, Tech. Rep. 11, Concrete Society, Slough, UK.

10. American Concrete Institute (1989) *In-place Methods for Determination of Strength of Concrete*, ACI 228 IR–89, ACI, Detroit.
11. Leshchinsky, A.M. (1992) Non-destructive testing of concrete strength: statistical control. *Materials and Structures*, **25** (146), 70–8.
12. British Standards Institution (1970) *Testing Concrete: Methods of Testing Hardened Concrete for Other Than Strength*. BS 1881 pt 5, BSI, London.
13. Basheer, P.A.M. (1993) A brief review of methods for measuring the permeation properties of concrete insitu. *Proceedings of the Institution of Civil Engineers: Structures and Buildings*, **99**, 74–83.
14. RILEM (1995) *Performance Criteria for Concrete Durability – State of the Art Report*. Rep. 12, International Union of Testing and Research Laboratories for Materials and Structures. E. & F.N. Spon, London.
15. Concrete Society (1988) *Permeability Testing of Site Concrete – A Review of Methods and Experience*. Tech. Rep. 31, Concrete Society, Slough, UK.
16. Parrott, L.J. (1994) Design for avoiding damage due to carbonation induced corrosion, in *Proceedings of the Third International Conference on Durability of Concrete*, ACI SP–145, American Concrete Institute, Detroit, pp. 283–98.
17. British Standards Institution (1988) *Testing Concrete: Recommendations on the Use of Electromagnetic Covermeters*. BS 1881 pt 204, BSI, London.
18. Bungey, J.H., Millard, S.G. and Shaw, M.R. (1994) The influence of reinforcing steel on radar surveys of concrete structures. *Construction and Building Materials*, **8** (2), 119–26.

19. Vassie, P.R. (1991) The half-cell potential method of locating corroding reinforcement in concrete structures. *Application Guide 9*, Transport and Road Research Laboratory, Crowthorne, UK.
20. American Society of Testing and Materials (1987) *Half-cell Potentials of Uncoated Reinforcing Steel in Concrete*. Designation C876–87, ASTM, Philadelphia.
21. Millard, S.G. (1991) Reinforced concrete resistivity measurement techniques. *Proceedings of the Institution of Civil Engineers, Part 2*, **91**, 71–88.
22. Gowers, K.R., Millard, S.G. and Bungey, J.H. (1993) The influence of environmental conditions upon the measurement of concrete resistivity for the assessment of corrosion durability, in *Proceedings of International Conference on Non-destructive Testing in Civil Engineering*, Vol. 2, British Institute of Non-destructive Testing, Northampton, UK, pp. 633–57.
23. Millard, S.G. (1993) Corrosion rate measurement of in-situ reinforced concrete structures. *Proceedings of Civil Engineers: Structures and Buildings*, **99**, 84–88.
24. Sadegzadeh, M. and Kettle R. (1986) Indirect and non-destructive methods for assessing abrasion resistance of concrete. *Magazine of Concrete Research*, **88** (137), 183–90.
25. Chaplin, R.G. (1990) *The Influence of GGBS and PFA Additions and Other Factors on the Abrasion Resistance of Industrial Concrete Floors*. British Cement Association, Slough, UK.
26. Moss, R.M. and Matthews, S.L. (1995) In-service structural monitoring – a state of the art review. *The Structural Engineer*, **73** (2), 23–31.
27. Carino N.J. (1994) Non-destructive testing of concrete – history and challenges in *Concrete Technology Past, Present and Future*, ACI SP–144 American Concrete Institute, Detroit, pp. 623–78.
28. Cheng, C. and Sansalone, M. (1993) Effect on impact-echo signals caused by steel reinforcing bars and voids around bars. American Concrete Institute, *Materials Journal*, **90** (5), 421–434.
29. Carino, N.J. and Sansalone, M. (1992) Detection of voids in grouted ducts using the impact-echo method. American Concrete Institute, *Materials Journal*, **89** (3), 296–303.
30. Pratt, D. and Sansalone, M. (1992) Impact-echo signal interpretation using artificial intelligence. American Concrete Institute, *Materials Journal*, **89** (2), 178–87.
31. Bungey, J.H. and Millard, S.G. (1993) Radar inspection of structures. *Proceedings of the Institution of Civil Engineers: Structures and Buildings*, **99**, 173–86.
32. Parker, D. (1994) X-rated video. *New Civil Engineer*, April 21, pp. 8–9.
33. Kear, P. and Leeming, M. (1994) Radiographic inspection of post-tensioned concrete bridges. *Insight*, **36** (7), 507–10.
34. RILEM (1994) Draft recommendation for damage classification of concrete structures. International Union of Testing and Research Laboratories for Materials and Structures. *Materials and Structures*, **27**, 362–69.

3

Achieving durable concrete

by J.G.M. Wood

ABSTRACT

Owners need to know comparative lifetime costs to select appropriate designs and specifications for new structures. They also require deterioration predictions for planning the cost-effective maintenance of their existing stock of concrete structures. We must therefore develop the means for predicting durability performance for major structures over a century or two, as society cannot afford the cost and disruption of replacing its built environment more frequently.

To achieve cost-effective durability design for concrete, we must combine a fundamental understanding and modelling of deterioration processes, derived from short-term laboratory studies, with long-term data from field structures with good and bad performance. We will need to change standards, design practice, materials specifications, contract procedures and site practice to achieve reliable cost-effective durability.

Keywords: Concrete, durability, design life, predictive modelling.

3.1 INTRODUCTION

Most of the design effort on concrete structures is concentrated on structural analysis to refine strength to achieve a structural form with least first cost. The specification for materials is usually a mixture of British Standards and 'cut and paste' clauses from previous contracts, recycled with little thought. The contractor concentrates his energy on maximizing his profit by constructing as fast as possible, with the cheapest materials which just comply with the minimum standards in the specification. The spalling rusty consequences of this approach are to be seen developing everywhere.

The home of the DoE at Marsham Street is having to be reconstructed after about 30 years because of premature deterioration of concrete. St

Christopher House where the DoT Highways Agency resides is little better. Let us hope that the message is getting through to the Government as the major client of the construction industry and with its national responsibility for standards. The DoE's *Durability by Intent: Strategy for Durability of Concrete* [1] is a small step forward. Those who queue on the motorways, as the frequency of repairs relentlessly increases, will be asking if the DoT is doing enough. The majority of the railways infrastructure was built of masonry and steel over one hundred years ago, and still remains serviceable and maintainable. Many of the motorway network concrete structures are likely to require major reconstruction every 40 years, if we extrapolate current trends. What is quite clear is that the minor changes in standards [2] and practice over the last decade have not begun to solve the long-term problem.

If we are to remedy this situation we will need to do more than wait for Eurocodes, in the hope that they will do better. The reality is that Eurocodes can be agreed only when they are close to the lowest standards of current practice in Europe and when they accord with the best interests of the material supply companies who dominate the committee structure directly or indirectly.

3.2 APPLYING A STRUCTURAL APPROACH

Structural engineering has evolved from an empirical art based on rudimentary calculations a century ago. From time to time major failures (Ronan Point, box girders, etc.) have led to a radical updating of standards. It is now a reasonably precise science in which the prediction of strength is based on analytical models which reflect the physical processes and which have been calibrated by testing large numbers of representative elements to failure. In the last 25 years partial factor design has been adopted with an associated refinement of loading and strength definitions. Factors of safety are based on the quantified variability of load and strength and the required reliability [3]. This approach has achieved economy and a reliability which has made structural failures exceptionally rare in the UK.

The same quantitative approach [4, 5] can be used to model the processes of concrete deterioration. The definition of environmental 'loads or actions' and material 'resistances' with factors of safety can be developed to achieve a similar reliability for the durability of concrete structures. The 'design life' [6, 7] and reliability required will vary from the needs of industrial sheds to those of nuclear facilities and critical elements of the transport infrastructure. Quantitative durability design, calibrated against field performance, will provide a rational basis for upgrading the overall design of structures. However, we must also critically review [8] and upgrade our models for lifetime costing. The hindsight evaluation of the costs of disruption and remedial work on structures like the Kingston Bridge in Glasgow and the Midland Links clearly demonstrates the folly of the fashionable 'build it cheap, for in 30 years there will be lots of money for repairs' approach.

Being able to predict durability will be of little value if we cannot improve it. The fragmentation of both the industry and the academic specialities relating to design and materials behaviour make it difficult to achieve change. We must find ways of improving communications between all those who influence durability. We must tighten the specification and contractual requirements so that quality can be defined and controlled in the finished structure. Building for quality must also be more profitable for the contractor. All of those in the concrete industry must contribute to the improvement of durability, as no single part of the industry can solve the problem. Durability can fail if there is just one weak link in the chain from the client's definition of design objectives, through the overall and detailed design, the materials selection and quality control, shuttering, steel fixing, compaction and curing, to the maintenance of water shedding from the structure.

The rate at which we can afford to replace our infrastructure necessitates design lives without maintenance for concrete structures of 100–200 years. This is easy where there are no chlorides and carbonation is the main problem. To achieve it for structures exposed to chlorides is a major technological challenge.

3.3 MATERIALS SCIENCE

Materials science has substantially improved our knowledge of the fundamentals of concrete deterioration from carbonation induced corrosion, chloride induced corrosion, sulphate attack, alkali–aggregate reaction, frost, etc. Many areas need further study, but the priority must be to relate laboratory work to construction practice. Some of our materials knowledge has led to improvements in codes, e.g. the banning of calcium chloride, and increases in cover requirements, etc. However, the strong prejudice in UK specifications against the use of admixtures to reduce water/cement ratios has left UK construction practice substantially behind that of the USA and Japan. We have also been slow to adopt slag and PFA for concrete exposed to chlorides.

Much materials science work on the durability of concrete is based on short-term laboratory testing in highly artificial conditions. Neville [9] has drawn attention to the academic constraints which have slowed progress in concrete technology. Durability studies in particular need a longer timespan and funding cycle than the three-year Ph.D. and Engineering and Physical Sciences Research Council grant module. It would help if the emphasis was shifted from labcrete to taking materials science skills out to analyse in detail the performance of structures with 5–50 years exposure to the elements. Unless we can relate laboratory work to real performance, we will be unable to reliably predict how materials will behave for periods of 100 years or more.

Fortunately there is now a growing body of longer-term data from exposure of specimens at the BRE marine site [10] and other long-term exposure sites. Traditionally these samples were assessed by taking chloride profiles in 10 or 25 mm layers and doing a simple Fick's law fit to two or three steps of the crude histogram. This blurred picture requires a very long period of exposure before clear trends emerge, particularly with the better PFA and slag cement mixes. If concrete needs a resistance sufficient to limit chlorides to the outer 40 mm after 100 years, it needs more precise analysis to measure diffusion rates. By re-analysing exposure samples in 1 or 2 mm layers, a clearer but more complex picture emerges (Fig. 3.1).

3.4 VINTAGE REALCRETE

Over the last 15 years I have been responsible for the investigation of concrete deterioration in several hundred structures worldwide. The data from these investigations is usually obtained for very specific purposes, so that decisions can be taken on the management of the structures. In consequence, the spread of parameters and, sometimes, the precision of data obtained is not always as good as in a laboratory

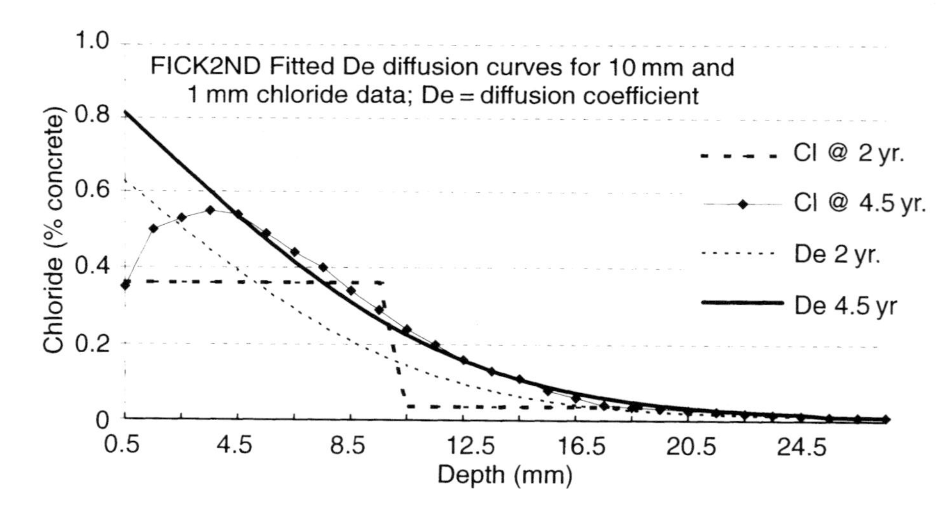

Figure 3.1 Analysis of 10 mm and 1 mm chloride ingress with Fick's law fitted profiles.

research project. However, the timescale of exposure of the structures and the quantity of data available from major investigations enables us to build a much better understanding of deterioration and the way it relates to the construction process. It also enables the variability of as-built materials performance to be quantified.

In some cases improvements in test procedures have been developed in the investigation of problem structures and in developing better specifications. Often these studies have been carried out with a substantial involvement by universities and research establishments (notably the Universities of Aston, Birmingham, Bristol, Wales at Cardiff, Dundee, Leeds and Sheffield, Imperial College, Queen Mary and Westfield College, BCA and BRE). The innovations have often been further developed in follow-on research programmes.

Figure 3.2 Tay Road Bridge. (Reproduced from Wood, J.G.M. and Crerar, J. Analysis of chloride ingress variability and prediction of long term deterioration: a review of data for the Tay Road Bridge; 1995).

The ability of the major commercial test laboratories to carry out extensive sampling and analysis on structures during investigation and repair contracts provides a data source which could be more widely used by the research community. Owners would substantially benefit from the proper analysis of this data to upgrade their design and specification requirements, but this is seldom done.

An example of the data which can be obtained and used to calibrate and refine models of deterioration processes is that from the Tay Road Bridge columns [11] (Fig. 3.2). This has greatly added to our knowledge of the change in chloride ingress with height above sea level (Fig. 3.3), and the variability of ingress which occurs in real structures due to uneven compaction and segregation (Fig. 3.4). Reducing variability 'up to the best' will be an essential factor in improving durability performance.

A recurring feature in the investigation of problem structures is the occurrence of localized very severe deterioration when detailing creates an adverse microclimate and/or poor cover and compaction. Often deterioration rates are ten times greater in these areas. They are particularly worrying where they coincide with areas of high stress, as in half-joints. Eliminating these features is an important element in durability design.

Long-term expansion tests on cores from structures (Fig. 3.5) and

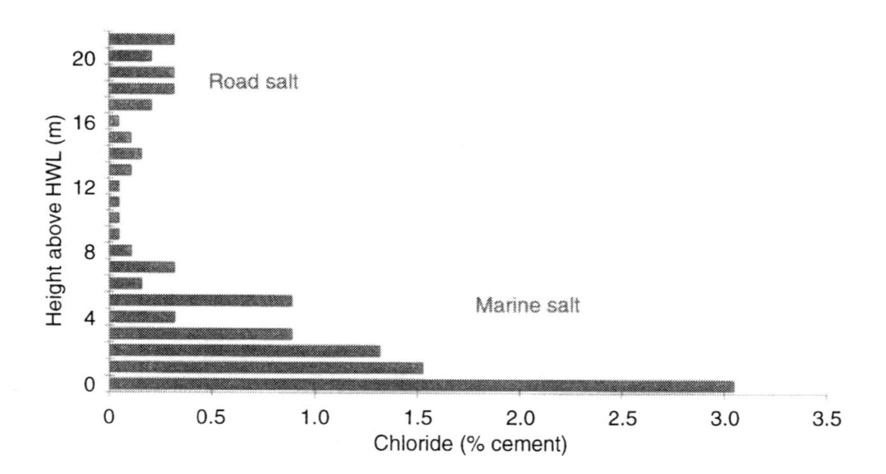

Figure 3.3 Variation of 0–25 mm chlorides with height up Tay Road Bridge Columns. (Reproduced from Wood, J.G.M. and Crerar, J. Analysis of chloride ingress variability and prediction of long term deterioration: a review of data for the Tay Road Bridge; 1995).

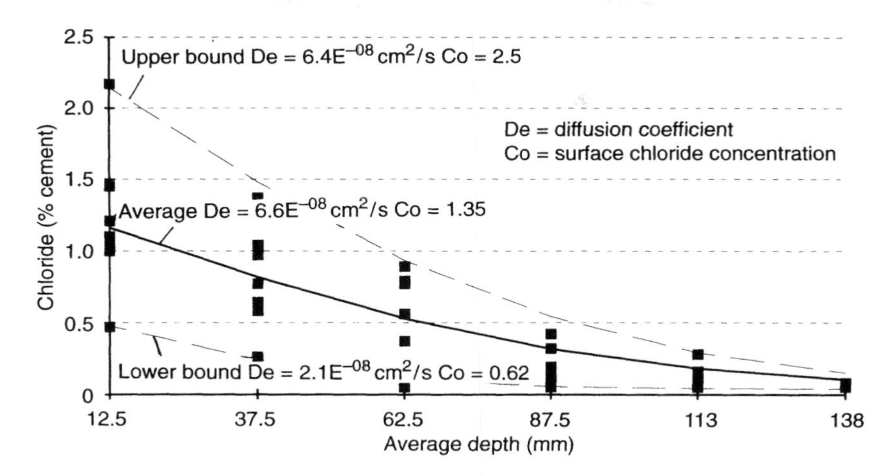

Figure 3.4 Variation of chloride ingress on the same face of five columns, at HWL + 1 m. (Reproduced from Wood, J.G.M. and Crerar, J. Analysis of chloride ingress variability prediction of long-term deterioration: a review of data for the Tay Road Bridge; 1995).

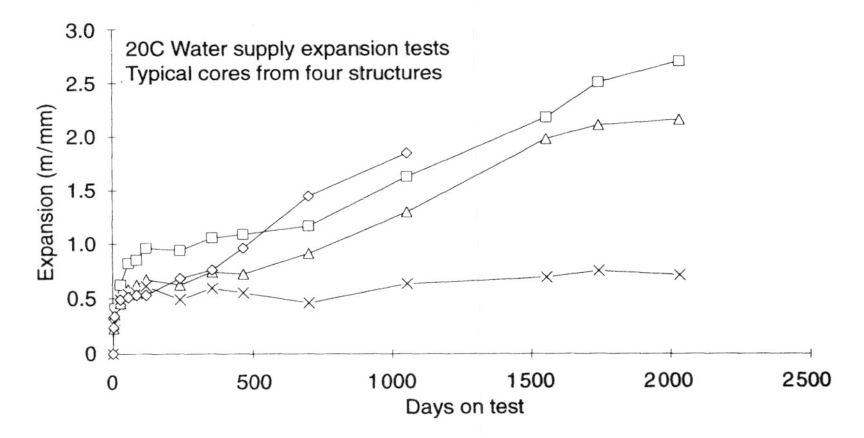

Figure 3.5 Long-term core expansion with AAR. (Reproduced from Wood, J.G.M. and Johnson, R.A. Alkali aggregate reaction in the United Kingdom: assessment of structural damage, remedial measures and the development of specifications; 1995).

trends of monitoring crack growth (Fig. 3.6) are giving us a much better picture of the timescale and variability of alkali–aggregate reaction (AAR) in structural concretes [12]. The field data comprehensively refutes the widely promoted view that AAR stops after ten years. The case of Montrose Bridge (Fig. 3.7) demonstrates that cracking can start to show after only 30 years and still be actively damaging the structure after 65 years [13].

Test programmes carried out to evaluate concrete for major new bridge and tunnel structures [14, 15] have greatly added to our knowledge of production quality concretes. The bulk diffusion test has been specified by Mott MacDonald to evaluate the chloride ingress resistance of different mixes on several major international contracts. This has provided 1 mm increment chloride ingress profiles for thousands of samples, which have greatly enhanced our knowledge of different cementitious materials in construction mixes.

3.5 MODELLING DETERIORATION BEHAVIOUR

Investigations of the condition of structures provide valuable data for the development and calibration of the predictive computer modelling of durability performance. However, the derivation of the equations comes from fundamental material science research worldwide.

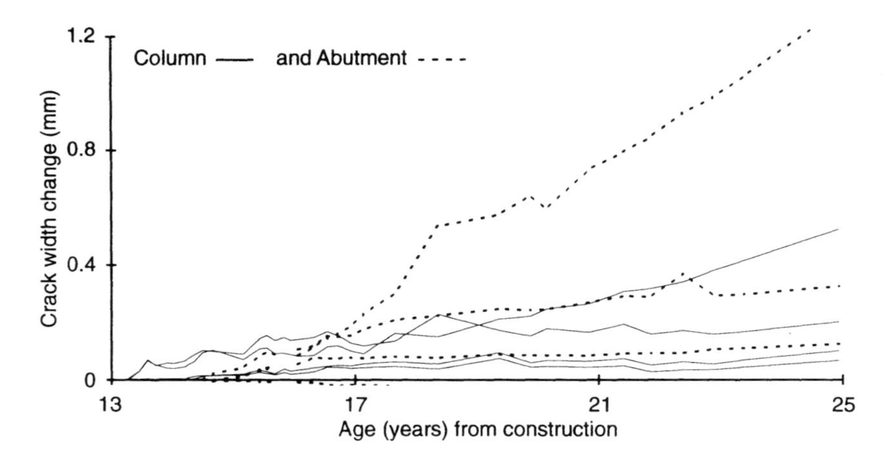

Figure 3.6 Long-term crack growth with AAR. (Reproduced from Wood, J.G.M. and Johnson, R.A. Alkali aggregate reaction in the United Kingdom: assessment of structural damage, remedial measures and the development of specifications; 1995).

Figure 3.7 Montrose Bridge. (Reproduced from Wood, J.G.M. and Angus, E.C., Montrose Bridge; 1995).

Simple Fick's law theory [16] helped develop our understanding of chloride ingress in the 1980s. It is sufficient for crudely extrapolating field data on chloride ingress from pure diffusion in OPC concretes, but has serious limitations for comparisons between different cementitious materials (PFA, slag, silica fume, etc.) and when exposure is more complex. Like all calculus solutions it requires extremely simplified assumptions, which limit the validity of the results when applied to the complexities of real concrete.

Finite-element structural analysis programmes enabled designers to break free in the 1960s from the constraints of the simplified calculus solutions of beam and truss behaviour. Now one can analyse complex highly redundant three-dimensional non-linear structures with plasticity, creep and buckling and dynamic behaviour modelled element by element.

The same approach can be applied to building non-linear deterioration models. These computer programs, like **CL$^-$IN** [5], can include elements which reflect changes in diffusion and binding with time. The concrete is modelled as a series of thin laminae which mirror the lamina-by-lamina analysis in Fig. 3.1. By modelling the flow of water and ions between the laminae over successive time steps, we can follow all the processes of hydration, binding of chlorides, surface carbonation, wetting and drying, changing diffusion resistance, etc.

Many of the physical and chemical equations that govern deterioration processes are well established in the literature. However, their application indicates that there are many interactions which need to be more fully evaluated. For example, when low water/cement ratios (> 0.4) are used the hydration desiccates the interior of the concrete to create a 'hydration suck' which will draw in pore solution with chlorides in addition to the diffusion migration of ions. Temperature effects are complicated by the changes in hydration rates and hydration products and binding, as well as the normal physical effects of temperature on diffusion in inert materials.

There is substantial further work to be done to develop the modelling of deterioration. There need to be associated reliability studies on the effects of the variability of as-constructed concrete on the risks of deterioration. A similar modelling approach can be used to develop more reliable specifications for other important deterioration processes like carbonation, sulphate attack, AAR [17], frost damage, etc. One of the exciting applications of this approach will be the development of improved cementitious materials tailored to the construction process instead of relying on ordinary cement.

3.6 CONCLUSIONS

The poor durability performance of many concrete structures is causing disruption and expenditure on remedial works which owners and society cannot afford and do not wish to see repeated. Piecemeal tinkering with current strength-based durability specifications will not achieve the major improvement in as-built durability required for concrete structures.

Durability design must abandon crude empiricism and follow structural design in developing reliability-based codes, in which variability and uncertainties are quantified and adequate factors of safety are included. We need and can achieve durabilities of centuries for major concrete structures, similar to those for traditional construction in stone, brick, timber and steel.

Fundamental research is leading to the numerical modelling of the detailed physical and chemical processes which govern material degradation. These models must be developed and calibrated using field data on vintage realcrete, which we can now more rigorously analyse to obtain statistically valid results.

Data from structures shows that materials' improvements will not achieve good durability against chlorides, unless the detailing is improved to ameliorate microclimates and to facilitate compaction, so that the variability of as-cast concrete is substantially reduced.

Using numerical modelling with site validated data, we can produce new, simple but reliable specifications for routine construction and a basis for evaluating the enhanced durability required for major projects.

Quantitative durability design must become a substantial and properly funded part of the design, specification and site supervision of new construction to ensure that overall design, detailing and materials selection achieve the requirements of owners and society. The marginal cost of enhanced durability is small compared to the long-term benefits.

Predicting future durability behaviour will also enable us to improve our management of existing structures which are deteriorating.

The construction industry must change its priorities from making concrete cheaply and quickly, to making it durable.

REFERENCES

1. Department of Environment (1994) *Durability by Intent: Strategy for Durability of Concrete*, November, DoE, London.
2. BS 8110 pt 1 (1985) *Structural Use of Concrete: Code of Practice for Design and Construction*, British Standards Institution, London.
3. CIRIA Report 63 (1977) *Rationalisation of Safety and Serviceability Factors in Structural Codes*, Construction Industry Research and Information Association, London.
4. Wood, J.G.M. (1994) Towards quantified durability design for concrete, in *Improving Civil Engineering Structures – Old and New*, (ed. W.J. French), Geotechnical Publishing Ltd, Basildon, UK, pp. 139–59.
5. Wood, J.G.M. (1994) Quantifying and modelling concrete durability performance. Paper to BRE meeting, November 10.
6. Somerville, G. (ed.) (1992) *The Design Life of Structures*, Blackie, London.
7. BS 7543 (1992) *Guide to Durability of Buildings*, British Standards Institution, London.
8. Gerwick, B.C. (1994) The economic aspects of durability – how much added expense can be justified? *Proceedings of Symposium on Durability of Concrete*, (ed. K.H. Khayat), American Concrete Institute, Nice, France.
9. Neville, A. (1995) Is concrete research likely to improve concrete? American Concrete Institute, *Concrete International*, May.
10. Thomas, M.D.A. (1991) Marine Performance of PFA Concretes. *Magazine of Concrete Research*, **43** (156), 171–85.
11. Wood, J.G.M. and Crerar, J. (1995) Analysis of chloride ingress variability and prediction of long-term deterioration: a review of data for the Tay Road Bridge. *Proceedings of Structural Faults and Repair – 95 Conference*, vol. 1, London.
12. Wood, J.G.M. and Johnson, R.A. (1995) Alkali aggregate reaction in the United Kingdom: assessment of structural damage, remedial measures and the development of specifications, in *Proceedings of International Symposium on New Developments in Concrete Science and Technology*, September 12–14, 1995, Nanjing. South Eastern University Press, Nanjing, China.
13. Wood, J.G.M. and Angus, E.C. (1995) Montrose Bridge, *Proceedings of Structural Faults and Repair – 95 Conference*, vol. 1, London.
14. Wood, J.G.M., Wilson, J.R. and Leek, D.S. (1989) Improved testing for

chloride ingress resistance of concretes and relation of results to calculated behaviour, in *Proceedings of Third International Conference on Deterioration of Reinforced Concrete in the Arabian Gulf*, October, BSE and CIRIA.
15. Leek, D.S. *et al.* (1994) Chloride ingress testing for high quality structural concretes in *Corrosion and Corrosion Protection of Steel in Concrete* (ed. R.N. Swamy), Sheffield Academic Press, Sheffield, UK, pp. 513–24.
16. Wood, J.G.M. (1995) Fitting chloride ingress data to a Fick's 2nd law diffusion curve, in *FICK2ND User Manual: Structural Studies and Design*, Guildford, UK.
17. Wood, J.G.M. (1993) Some overseas experience of alkali aggregate reaction and its prevention – specification for major projects: bridges, tunnels and dams. Technical Institute of Buildings and Public Works. *Paris ITBTP Annales*, (518) November.

Discussion of Chapters 2 and 3

Dr Adam Neville If we try to answer, before we start proper discussion, Dr Wood's question 'What did the universities do wrong when they forecast?', one answer is this – that there exist two excellent structural materials: one is steel (if you forgive me), the other one is concrete. You only go wrong when you put the two together. I am not being entirely jocular; in one of the areas to which Dr Wood referred there have been built some very major structures using reinforced concrete and yet the same structures could have been built using plain concrete, just as they used to be built using stone, and it is the steel that does it. Now, that is not meant to be an unhelpful remark but it requires alternative thinking in terms of the type of structure that we build.

Well, I apologize for this excursion. We now have time for discussion on all three papers presented this morning. Who would like to counter-attack Dr Wood?

Professor D.J. Hannant, *Department of Civil Engineering, University of Surrey, Guildford* Over the past 25 years I have been teaching concrete technology to students and explaining how to make concrete durable. Every year or two I teach different rules because the eminent engineers who sit on committees make new rules every few years which are often quite different from the previous ones. I am sure that if the experts in this room today sit down and write a new set of guidelines as to how to make concrete structures last 50 or 100 years, and we wait 10 or 20 years to see how the structures perform, the guidelines would again be substantially changed. So, the problem is that we are hitting a moving target; materials are changing all the time and we have to wait 20 or 30 years to find out whether our ideas are correct. I venture to suggest that everybody in this room will be dead before we find out whether today's rules are actually correct – that is 70 years hence. So we are on to a bit of a loser. We do the best we can with the knowledge we have but because academics do not, on the whole, write codes of practice, they can hardly be held responsible for the lack of durability of structures.

Dr Jonathan Wood On Engineering and Physical Sciences Research Council committees we get applications from universities saying 'We are experts on this and want to do more research, which will help the committees developing the rules.' Thus many universities are in the forefront of the development of improved guidance, rather than waiting to be told what to tell their students. We must also acknowledge that the flow of information from the research community into standards and guidance has been seriously disrupted and delayed by the diversion of effort from British Standards into Eurocodes. These are often agreed at the lowest current standard, rather than moving forward to raise standards.

Predicting 20 to 25 years ahead is getting a lot easier as our ability to analyse concrete is improving. By using scanning electron microscope/ electro-dispersive X-ray analysis or millimetre by millimetre profile grinding to analyse in great detail the surface few millimetres of concrete, we can identify and quantify the initiation of deterioration processes. This precise data from a year or two of site exposure, or a few months' accelerated laboratory exposure, can be extrapolated to predict future behaviour. We no longer need to wait a decade or two to see what develops. Great advances have also been made in the petrographic examination of concrete.

Many of the potential innovations in UK concrete practice have long track records in other countries, e.g. the use of silica fume in Norway and Iceland and the use of slag cements in the Netherlands. By rigorously examining quantitative analytical data on their performance, one can form a good picture of their likely performance here. However, some of the anecdotal reports in sales literature need sceptical evaluation.

One of the hopes for the next 20 years is that we will move on from considering cement as just an 'ordinary' grey powder with broadly defined characteristics. Potentially we can get 'special' cementitious materials which are formulated specifically to give enhanced durability properties and to facilitate construction quality and consistency.

Dr Adam Neville If I could add to that, I wonder whether universities should teach rules – I think they should teach understanding of what happens and if you have got that understanding you can apply it to any material once you have got further experience somewhere else.

Graham West, *retired (formerly Transport Research Laboratory)* I would like to make a comment on an earlier remark that Mike Grantham made concerning the bringing-in of chlorides to concrete multi-storey car parks by means of the parking in them of vehicles whose tyres are wet with de-icing salt in the winter: the salts build up in the car park and do not readily get out. There is another situation where this happens and that is in concrete-lined road tunnels. Here it is not just a question of cars being parked, but the continual passage of vehicles bringing salt in to an environment where there is no rain to wash it out. So this is another situation where a possibly deleterious concentration of chlorides can occur in a concrete structure.

Dr Jonathan Wood This is not just a possibility, it is a reality in many road tunnels, as well as car parks and bridge soffits. The importance of rain wash in removing chlorides is often underestimated. Where rain washes chlorides off structures it prevents a surface concentration build-up. Without rain wash this can reach 10 to 15 times that of normal sea water exposure, with a corresponding increase in chloride ingress rates.

Dr Adam Neville The question of chloride ingress in structures of the type considered brings us back, if I may add a few words, to the composition of the concrete. I think it was John Glanville who referred to the British reluctance to put admixtures into the concrete which can be contrasted with the American attitude where you formulate your mix. I disagree with Jonathan Wood; I could put a full stop here, but I disagree with him on the question of formulating special cement. I think cement, Portland cement, has to remain a relatively cheap material; I am using the word 'relatively', being conscious of the motor cars which cement industry people drive, but it is other materials that have to be used to produce the suitable mixture. Admixtures include waterproofing admixtures. There is a tremendous reluctance to use those just because the early ones were not lasting in the sense that they disintegrated chemically or underwent some other changes but when one talks about those things these days, the general reaction is 'Well, can you prove that the admixture is effective over a period of 30 years?'. With that attitude you don't get very far.

Dr Poole, *Queen Mary and Westfield College* I want just to bring us generally back to the problems raised earlier by John Glanville. I wonder whether one of the speakers or delegates who are experts in the field would like to speculate on what new sorts of problems may occur in the future with precast units when we consider concrete durability.

Dr Adam Neville Who is willing to speak on this?

Simon Fawcett, *Montgomery Watson* In connection with that last subject, we had a project overseas where I noticed that the use of superplasticizers changed the electrochemical characteristics of the concrete quite markedly and this is something which I have not been able to find any research on at all and I do not have the funding to undertake that kind of research.

Dr Adam Neville Anybody else on that topic?

Dr Jonathan Wood One of the new problem areas is very high-strength concrete (> 80 N/mm^2). The Victorians had very high-strength cast iron, but its brittleness led to many failures. They then realized that ductility was the essence of safe structures and moved to using lower-strength ductile mild steel.

These very high-strength concretes are very brittle and can lose strength with time as internal self-desiccation shrinks the cement paste off aggregate particles. The brittleness may invalidate normal structural analysis, which is based on a degree of redistribution of stress at stress concentrations and as ultimate loads are approached. The self-desiccation can also draw chlorides in. There are already problems developing with these materials and I expect more to come! The introduction of silica fume in concrete will also give some unexpected problems, particularly when it is not well dispersed.

Dr Adam Neville Well, I said two minutes ago Jonathan Wood and I disagree, full stop. I repeat this. Now, if silica fume is not well dispersed, then you should sack somebody; there is no difficulty in dispersal. Secondly, one avoids the shrinking off the aggregate by using aggregate of a maximum size of 10 mm. Thirdly, in his presentation, Jonathan Wood spoke about a lower water/cement ratio of 0.35 – well, this is where he gets trouble. If you use 0.3, 0.28, 0.25 then the system through which transport can take place, and I have not heard Nick Buenfeld yet, is such that there is not much danger of aggressive agents entering the concrete. There is a problem if there is a fire very early, when water can't get out, but I think writing off those materials is yet another example of not moving too fast just because the Canadians have done it, and not invented here. I know you disagree with me and we can take this as fact. I wonder whether John Bungey wants to respond to the dismissal of non-destructive testing – you can see I am looking for allies!

Professor John Bungey Yes, thank you Chairman for the opportunity to come back on this. In terms of the comment about non-destructive testing, I think that the comment is so flawed that if you follow Jonathan's approach you have no structure left to test or use, or do anything with. Obviously, the compromise has to lie somewhere in-between. I fully accept that to validate non-destructive testing some degree of excavation, to see what you really have got and to prove it, is necessary. The great potential value of non-destructive testing as far as I can see is, first of all, establishing where is the best place to dig those holes. This gives you extra guidance over and above what your eyes and common sense tell you. Secondly, it will also enable you to extend those findings to other parts of the structure to save you having to dig the whole thing up. Whether Jonathan will agree with me on that or not, I don't know, but that would be my response. I wonder if I might also, whilst I have the microphone in my hand, just refer to his implied criticism of thinking that everything from the USA is wonderful. We have to be very careful that we do not fall into the same trap as many engineers and researchers in the USA who consider that the world ends at the boundaries between the USA and the Atlantic or the Pacific. I think that one of the strengths of UK research is that it does go to great lengths to establish what is going on in other parts of the world and tries to learn from this. I think that there are, perhaps, some lessons to be learnt from their experience, whether it is successful or unsuccessful.

Dr Adam Neville Thank you.

Mike Walker, *Concrete Society* I thought John Bungey's outline of NDT was very good and I was interested to hear what he had to say. John is, amongst many other things, the chairman of the non-destructive testing working party of the Concrete Society and a member of the society's materials group. During his presentation he mentioned that the Concrete Society is going to publish a document on the relationship between durability and types of specification. This is so, but I would like to stress that this is to be in the form of a discussion document. In it, performance specifications will be compared to prescription specifications, and the true role and place of testing, and the use of quality assurance in specification requirements and control will be considered. Its publication will provide a review of existing procedures and possible alternatives for the industry to consider and debate. It will not be a normal society technical report giving information and guidance or recommendations for immediate use. The discussion document status of this publication is important as there is sometimes a tendency for some sections of the industry to apply published proposals and suggestions without realizing that they are not necessarily recommendations. For instance, in the society's permeability report, *in situ* permeability tests were referred to and it was made quite clear that until the tests are developed further and there are sufficient results to relate measured values with real performance, such tests should not be used in specifications. Unfortunately there have been areas where this advice has been ignored and *in situ* permeability testing specifications have been developed using the values outlined in the report.

Dr Adam Neville Thank you very much, Mike, that was a really useful remark and I am sure we want to be aware of what is coming and be ready to discuss it.

John Glanville Bill Wolmuth and I went to the States earlier this year on a tour to study aspects of building technology, nothing to do with reinforced concrete. It was, however, extremely illuminating because we met people from test laboratories, engineers' and architects' offices, and made a number of site inspections; our overwhelming conclusion was that they were some years behind us. Now, I am sure there are areas where they are ahead of us, but in that particular field we came to the conclusion that European practice, both in design and in construction, was far in advance.

Dr Adam Neville Thank you. One more contribution.

Mike Grantham, *MG Associates* Just going back on what John (Glanville) says, I wish I could agree with you, John. You are probably right in some areas but I think in some areas the States are way ahead of us. As part of a legal job I am involved in at the moment, I had to do a review of what current practice was with car park design in the years 1980–8 to see how we stood and how things were internationally. In about 1980 in the States they were beginning to sound warnings about car parks and de-icing salts. By about 1984 they were saying the evidence is strongly building about car parks and de-icing salts, by 1988 they were saying they should be waterproofed; in 1988 in this country engineers were publishing papers which said 'the problems with car parks relate to cracks and the cracks allowing moisture through, which causes staining on the cars underneath'. They simply had not addressed the issue of chlorides at all in this country and, as I say, I think we still have the legacy of this today in that the state of knowledge is still pretty limited on this subject; so there are undoubtedly areas I think where we are ahead of them, but you know, the same applies with them as us and I think it is very important to look beyond your national boundary when you are trying to get information on durability-related aspects.

Colin Gallani, *C.L. Gallani Associates* One aspect that has not been brought out thus far is that durability of concrete at the end of the day is related to the way in which the concrete structure is actually built and therefore to the design of the formwork, design of the reinforcement, the way in which the formwork is erected, the way in which the steel is fixed and the way in which the concrete is poured. When specifying concrete, therefore, these aspects must be considered. Further, one has to bear in mind that the skills generally involved in the construction functions on sites are relatively low-technology skills and that these impose practical limitations on the achievement of concrete to an advanced or sophisticated specification.

Dr Adam Neville There is another discussion period at the end of the afternoon and I think we might be able to arrange to cover some of the morning subjects as well. In the meantime, there is one other function which I have to perform and that is the presentation of the STATS prize. STATS has generously established a prize – I think in 1985 – so it is just ten years ago, which is awarded to a student who has completed the M.Sc. course in Geomaterials at Queen Mary and Westfield College. The prize is awarded, although it is a STATS prize, on the basis of a recommendation from the Board of Examiners. It is a pure coincidence that I was an external examiner. It is an annual prize but it is not awarded every year, for example it wasn't awarded last year, because it is only given when particular distinction is achieved in the dissertation. Now, this year's prize is going to be presented to Denise Griffiths-Richards who submitted a dissertation with a very long title: 'Factors affecting the hydraulic efficiency of granular drainage layers in leachate collection systems of engineered landfills'. She received her degree of M.Sc. with distinction and, in addition, she is now going to receive the prize. The prize is in two parts, one part is this volume and I believe there is part two coming but the book which she is to receive is still being printed. It gives me much pleasure to present the prize to Denise Griffiths-Richards.

That completes the morning's proceedings.

4

Engineering design and service life: a framework for the future

by G. Somerville

ABSTRACT

Some ten years ago, the author [1] made a distinction between the production and placing of durable concrete, and the design and construction of structures that are durable. The essence of the argument was that the industry tried to solve all its durability problems via a prescriptive materials approach, and yet standards of design, detailing and construction were at least as significant.

In the meantime, a number of relevant and important trends can be noted:

- a growing interest, on the part of owners, in life-cycle costing and in the need for planned management and maintenance – both for the assessment of existing structures and the design of new-build;
- the development of a perspective of the relative importance of the different deterioration mechanisms, and of predictive models to quantify their effects;
- significant improvement in the ability to specify durable concretes for specific aggressive actions.

In the light of these trends, this chapter reconsiders the proposals made ten years ago on a design life approach, and suggests a framework for future design which embraces the new technologies to meet owners' defined performance requirements. In particular, a quantified approach to design for durability is proposed – similar to, and integral with, conventional structural design.

Keywords: Design, durability, engineering framework, performance criteria.

4.1 INTRODUCTION

The title of this book is *Prediction of Concrete Durability*. The aspect addressed in this chapter is how performance prediction can contribute to engineering solutions in each of two situations:

- the design of new structures to provide satisfactory performance in terms of strength, stiffness, stability and serviceability with time (durability) at minimum cost (increasingly, this means whole-life cost);
- the assessment of existing structures, in terms of current state and future deterioration, and hence of residual useful life.

Neither design nor assessment is ever likely to be an exact science. Engineering judgment and experience will always be required. Hence the approach adopted here will be to explore how the underlying understanding of concrete, the material, coupled with the science of prediction and modelling can best contribute to these real-world engineering applications. The approach will, unashamedly, be an engineering one.

Some ten years ago, the author [1] made a distinction between the production and placing of durable concrete, and the design and construction of structures that are durable. The essence of the argument was that the industry tried to solve all its durability problems via a prescriptive materials approach and yet standards of design, detailing and construction were at least as important.

Put another way, even the best of concretes, correctly specified and placed, can give poor performance in a bad conceptional design associated with poor detailing. This is especially true in allowing for movement and in dealing with water. Two examples to make the point: Fig. 4.1 shows the effect of joints on the in-service performance of 200 bridges. Figure 4.2 shows possible mechanisms by which rainwater can leak through the cladding of buildings. How many of these are known to the typical designer when detailing the joints?

Engineering design will be covered in detail later. It must first be said that there are some aspects of durability which are best handled by a prescriptive approach. Mostly, these are aggressive actions which affect the concrete directly. A selection is given in Table 4.1, which also summarizes the general approach. Enough knowledge exists to deal with each of these individually, although compromises have to be made in practice, when more than one such action can occur. There are also moves afoot to develop performance-based approaches to freeze–thaw action and to abrasion; as always with this approach, the difficulty lies in agreeing and developing a performance-based test, which has the simultaneous attributes of being representative of real conditions and of being practical under contractual conditions in precision terms.

Picking off individual aggressive actions and dealing with them in this way has its attractions. However, it does raise the question of the objectives of the more general durability provisions which appear in codes and standards. On a worldwide basis, these provisions can vary significantly in detail, but have common features:

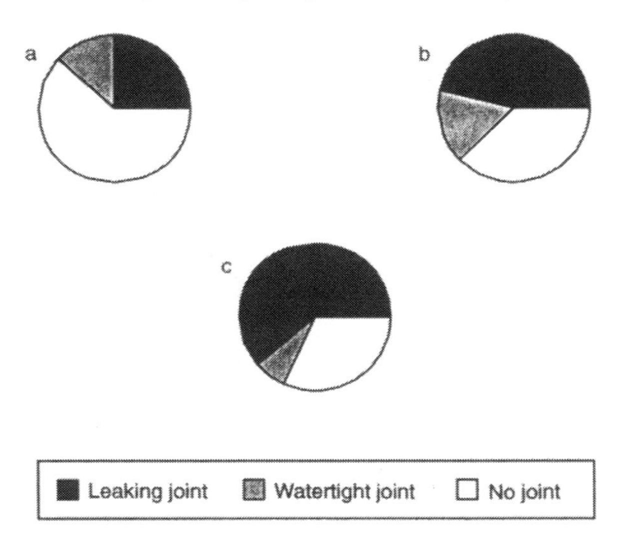

Figure 4.1 Influence of joints on bridge condition; a – good condition, b – fair condition, c – poor condition. (Reproduced from Wallbank, E.J., *Performance of Concrete in Bridges: A Survey of 200 Highway Bridges*; HMSO, 1989.)

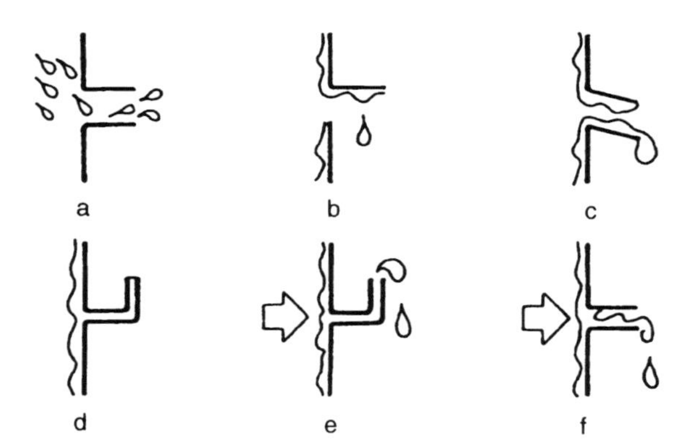

Figure 4.2 Possible mechanisms by which rainwater leaks through joints; a – kinetic energy, b – surface tension, c – gravity, d – capillarity, e – pressure assisted capillarity, f – air pressure differentials. (Reproduced from Anderson, J.M. and Gill, J.R., *Rainscreen Cladding: A Guide to Design Principles and Practice*; CIRIA, 1988).

Table 4.1 Types of aggressive action for which material specifications have been developed[a]

Aggressive action	*Comments*
Sulphate attack	Specific material and mix proportions are recommended in most codes for defined ranges of sulphate concentration.
Alkali–silica reaction	The basic reaction and its possible effects are now well understood. Recommendations to minimize the risk of damage are published.
Freezing and thawing	Dealt with by choice of materials, mix proportions and concrete grade. Air entrainment for lower grades. Detail to minimize exposure to moisture.
Abrasion	Specifications to cover aggregate properties, concrete grade and mix proportions, compaction and curing, methods of finishing, etc.

[a]General approach: (a) quantify the action (b) define ranges of intensity for it (c) produce a specification for each range.

- a classification of exposure conditions;
- an emphasis on achieving low permeability, translated in practice into detailed recommendations, in what has been described [1] as 'the four Cs' – constituents of the mix; cover; compaction and curing – for each exposure class.

The classification system of exposure tends to be couched in fairly loose general terms. Exposure conditions defined in BS 8110 [4] are shown in Table 4.2; examination reveals an underlying concern with moisture and leads to the possible conclusion that the main concern is with corrosion. This is confirmed by the follow-up prescriptions for mix proportions and cover, where the objective is to get a concrete of adequate quality and thickness surrounding the reinforcement. However, the only compliance check is for strength, measured on standard control specimens, and there is no real assurance that the desired end result of low permeability in the cover concrete is achieved.

One might reasonably expect in the 1990s that there would be unanimous agreement on specifying the four basic requirements (the four Cs) for any given exposure condition. This is not so. There are some remarkable differences in various national codes – apparently without corresponding differences in the durability actually achieved. The situation is further complicated by changes in concrete technology, particularly in the use of additives and additions, but also in a general shift towards what has become known as 'high-performance concrete'.

Table 4.2 Exposure conditions defined in BS 8110[a]

Environment	*Exposure conditions*
Mild	Concrete surfaces protected against weather or aggressive conditions
Moderate	Concrete surfaces sheltered from severe rain or freezing while wet Concrete subject to condensation Concrete surfaces continuously under water Concrete in contact with non-aggressive soils
Severe	Concrete surfaces exposed to severe rain, alternate wetting and drying or occasional freezing or severe condensation
Very severe	Concrete surfaces exposed to sea-water spray, de-icing salts (directly or indirectly), corrosive fumes or severely freezing conditions while wet
Extreme	Concrete surfaces exposed to abrasive action, e.g. sea water carrying solids, flowing water with pH < 4.5, machinery or vehicles

[a]Source: BS 8110, *Structural Use of Concrete*; published by BSI, 1985.

These differences and changes make learning from in-service feedback rather difficult. Whatever might have been learnt from the past has to be applied in a different technological world. More fundamentally, one might question the validity of the system: if corrosion is the prime durability issue, why not treat it directly as for other deterioration mechanisms? An understanding of cause and effect in the corrosion process is then essential, but it then becomes clear that the basic objective is the prevention of significant loss of rebar section – or, more generally, of resisting tensile force. This immediately opens the door to a whole series of different options to achieve that objective, beyond the normal option of trying to ensure that the cover concrete is largely impermeable.

This type of thinking, however simplistic at first sight, immediately moves durability considerations into a design mode – the subject of the rest of the chapter. However, before doing that, one final important point has to be made about current practice in construction technology. While we may now know much more about different types of concrete and might even succeed in translating this knowledge into appropriate specifications, all the signs are that we are still failing to ensure that these specifications are met in practice. The consistent achievement of the required cover is perhaps the best example of this. Building quality into our procedures and skills has to be a high priority, no matter how much we improve the design and materials side of things.

4.2 WHY AN ENGINEERING DESIGN APPROACH FOR DURABILITY?

Much of our concrete-based infrastructure has now been in place for several decades. There is some feedback on technical performance, at least in qualitative terms (Fig. 4.1); this demonstrates that concrete is not a maintenance-free material in general terms, but also that the level of deterioration in individual cases is dictated by a combination of factors in which design and construction issues are significant [2, 5]. The net effect can be a costly loss of function, or substantial unforeseen maintenance costs – or both.

However, the spectre of obsolescence also enters the arena. Many buildings and bridges have had to be upgraded or replaced, because their functional needs have changed – quite apart from any decrease in technical performance. In addition, different components in individual artefacts (e.g. cladding in buildings; expansion joints in bridges) have been shown to have useful lives much less than those for the basic structure.

These factors have led to a more conscious effort to manage and maintain the existing infrastructure, and to introduce life-cycle cost techniques to evaluate alternative designs for new structures, e.g. [6], or to develop performance profile plans for complete buildings as illustrated in Table 4.3. It may be seen from Table 4.3 that the concept of life is now entering into design thinking: indeed, this has been formalized with the publication of a helpful British Standard [8].

In the context of this chapter, all of this means that durability – an integral part of technical performance-in-time – has to be considered as part of the overall design process, in meeting these changing needs of owners, expressed in terms of function and finance. This remark applies equally to the assessment of existing structures. The question now is: what sort of design framework is required to achieve this, compatible with other design considerations?

4.3 ELEMENTS IN A DESIGN FRAMEWORK

There are really two phases in design:

1. initial design, where alternative schemes are compared in terms of functional and architectural suitability, on the basis of cost. From this emerges the favoured solution in terms of concept, if not detail. Decisions taken in this phase will have a fundamental effect on

Table 4.3 Basics of a performance profile plan for building[a]

System[d]	*Criticality*	*Target life before replacement, years*						*Capital costs*[b]	*Cost in use target: x years at y price*[c]
		>5	5–10	10–20	20–40	40–100	>100		
Foundations	A	●●●●●	●●●●●	●●●●●●	●●●●●●	●●●●●	●●●	–	–
Structure	A	●●●●●	●●●●●	●●●●●	●●●●●	●●●		–	–
External cladding	B	●●●●●	●●●●●	●●●●●	●●●●●			–	–
Glazing	B	●●●●●	●●●●●	●●●●●	●●●			–	–
Partitions	B	●●●●●	●●●●●	●●●●●	●●●			–	–
Heating and ventilation	B	●●●●●	●●●●●	●●●				–	–
Water, public health services	A/B	●●●●●	●●●●●	●●●●●	●●●●●			–	–
Electrical	A	●●●●●	●●●●●	●●●●●	●●●●●			–	–
Decorating	C	●●●●●	●●●●●					–	–

[a]Source: White, K.H., *The Structural Engineer*, **69** (7), 148–51; 1991.
[b]For example, 100 units.
[c]For example, 1500 units over x years.
[d]Maintenance requirements: clear specification required for each system.

performance in service, and should be based on experience and on the best available feedback.
2. final design, where the initial concept is checked out fully, and final working details produced.

For final design in particular, there are five essential elements in the process, in checking structural adequacy in quantitative terms. It is argued here that these same five elements could also be present in developing a quantitative approach to durability design. This is illustrated in Table 4.4, which lists the items for structural design and shows parallel (but integral) possibilities for durability design. This immediately shows where performance prediction fits in – the theme of this book – but also indicates that it is only one element of the package, where there has to be a balance between all five.

The key point about Table 4.4 is that it attempts a quantitative approach to durability. To be successful, durability design requires more detailed knowledge of item 1 than with structural design. However, the most significant items are probably 2 and 4, where little work has been done. With respect to item 2, one needs to know what one is trying to predict, since this might influence the predictive model to be used. Item 4 suggests that one needs to introduce the concept of safety, i.e. to set margins which reflect the precision of the process.

To understand better what is intended here, it is necessary to look at these elements in a little more detail, and at the relationship between them. A possible relationship is shown in Fig. 4.3. Each of these elements will now be treated in some depth.

4.3.1 The required performance cube

A proposal for this is shown in Fig. 4.4. The front face of the cube is used to make a statement about the importance of the structure or element, in terms of life and criticality. Whilst there has been discussion in the literature regarding specifying design lives for different types of structure in years [1, 8, 9], this is not considered necessary. The idea shown is based on a CEB concept, and the objective is to offer a series of zones (A–G in Fig. 4.4) which define required performance in broad qualitative terms. A possible grading for criticality might be:

- high: failure would cause cessation of function and/or major disruption during remedial work;
- medium: efficiency of operation would be reduced, but replacement/ remedial work could be done out of normal working hours;
- low: not critical. Any necessary maintenance or remedial work could be done without inconvenience.

Table 4.4 Elements in a design framework

Item	*Actual for structural design*	*Parallel possibilities for durability design*
1. Loads	Imposed loads taken from codes	Classification of environments Identification and quantification of aggressive actions
2. Performance criteria	Adequate strength, stiffness and serviceability Deflection and crack width limits	A statement of required life in qualitative or quantitative terms Some account of criticality (risk analysis) A definition of a performance profile including any strategy for maintenance Specific limits to 'damage' or effects of deterioration (e.g. cracking or loss of section due to corrosion; expansion due to ASR; internal damage due to freeze–thaw)
3. Modelling analysis	Methods of analysis used to determine action effects due to the applied loads Design equations used to provide resistance to the action effects (bending, shear, etc.) Recommendations on detailing	Predictive models to determine the effects of the aggressive actions Models/equations used to calculate the effects of deterioration on conventional action effects Evaluation of alternative options in providing the required resistance for the required time – usually a combination of material, design and construction options
4. Factors of safety, margins	Partial factors which are generally applied both to the loading and resistance sides of the design condition $S \leqslant R$ Can also be built into design equations Margins may be set, in establishing limiting performance criteria (e.g. crack widths)	Ideally, should require the same approach as for structural design Should be done consciously, depending on knowledge of loads, required life, risk analysis, precision of models, etc

Table 4.4 (*continued*)

Item	*Actual for structural design*	*Parallel possibilities for durability design*
5. Specifications, certifications, QA	Concrete mix ingredients and proportions Cover Rebar specifications Minimum workmanship requirements Supporting certification and QA schemes	Essentially the same as for structural design, but with more factors included (e.g. coatings, special steels, cathodic protection) and possibly more options (e.g. different classes of construction) Supporting certification and QA schemes may have to be stronger

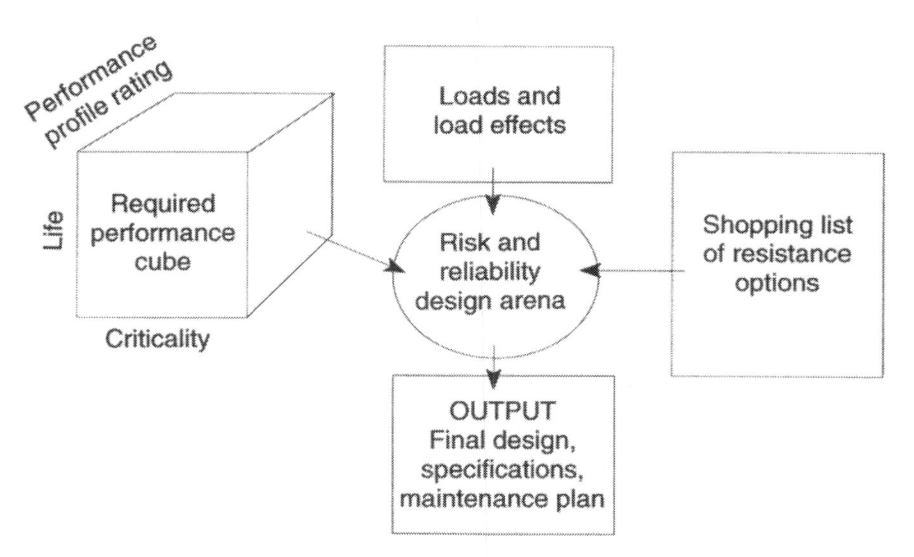

Figure 4.3 Suggested outline relationship between the elements in a durability design framework.

The grading is similar to the British Standard idea of classifying components as replaceable, maintainable or lifelong [8]. Any foundation or key structural element might be categorized as high criticality or lifelong, but again this could depend on the type of structure; different types have different lives.

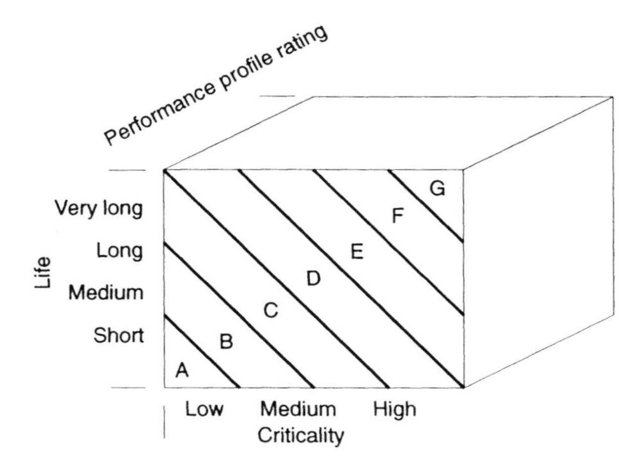

Figure 4.4 The required performance cube.

More significant, at the detailed level, is the third axis of the cube, labelled performance profile rating. There are two elements in this:

1. a clear statement on maintenance–replacement strategy, derived via life-cycle costing methods, and related to the perceived future functional requirements. This has a direct effect on the required performance profile (Table 4.3).
2. the establishment of limiting performance criteria for individual aggressive actions. These may be different, depending on the choice of zones A to G. Actions would be dealt with individually and collectively, and would include (say) corrosion, freeze–thaw, sulphates, chemical attack, alkali–silica reaction, abrasion and cracking.

As an example of what is intended here, consider corrosion. A list of possible limiting performance criteria might include:

1. The mean maximum level of a carbonation or chloride front should not penetrate more than (a)% of the nominal cover.
2. Corrosion has just started, i.e. the critical front has reached the reinforcement, both oxygen and water are available, and, in the case of chlorides, the critical chloride threshold level has been reached.
3. Corrosion has caused cracking parallel to the reinforcement and the crack width is equal to (b) mm.
4. Corrosion has removed (c)% of the cross-section from (d)% of the reinforcement.

This type of thinking is already advanced in assessment work; for example, the CEB has suggested five levels of damage classification as indicated in Table 4.5. The current proposal, covering both design and assessment, is no more than an extension of that.

Referring to Fig. 4.4, examples of limiting criteria which might then emerge are:

- zone A: mean maximum level of a chloride profile should not reach the rebar for 15 years; loss of rebar section (< 10%) after 30 years.
- zone G: mean maximum level of the chloride profile should not reach the rebar for 100 years.

These examples require very different design solutions.

The following notes to the proposal are relevant:

1. A knowledge of the maximum likely corrosion rate is always necessary, under the microclimate conditions likely to be present on site.
2. The ability, in modelling terms, to predict each of the suggested

Table 4.5 Damage levels of reinforced concrete due to steel corrosion[a]

Visual indications	*A*	*B*	*C*	*D*	*E*
Colour changes	Rust stains	Rust stains	Rust stains	Rust stains	Rust stains
Cracking	Some longitudinal	Several longitudinal Some on stirrups	Extensive	Extensive	Extensive
Spalling		Some	Extensive	In some areas steel is no more in contact with concrete	In some areas steel is no more in contact with concrete
Loss in steel section		5%	10%	25%	Some stirrups broken. Main bar buckled
Deflections				Possible	Apparent

[a]Source: Comité Euro-International du Bréton (CEB).

limiting criteria , 1–4 above, is essential. The ability to predict the time from one criterion to the next is desirable.

3. The possibility of permitting extensive cracking and loss of steel section, as proposed in Table 4.5 for assessment work, is not as daring as it might first appear. A 10% loss of section does not automatically mean 10% loss in load-carrying capacity. Moreover, for rates of corrosion which occur in very many cases in the field, it can take very many years to lose 10% of the cross-section of large-diameter bars.

Notes 1 and 2 perhaps reflect objectives for those involved in modelling deterioration processes. Note 3 is directed more at engineers and owners of structures, by indicating that, for reinforced concrete at least, there is usually sufficient time to evaluate alternative remedial strategies, compatible with longer-term functional needs, without significant risk of failure. An underlying need, in developing such an approach, is a rethink of current serviceability criteria, with a much greater emphasis on the physical effects of deterioration.

4.3.2 Shopping list of resistance options

In current activities aimed at improving durability standards for grouted post-tensioned bridges, Raiss [10] has proposed the concept of multi-layer protection, in consciously designing for durability. In effect, this recognizes that any single method of protection may fail or wear out, and there is a need for redundancy in the design system. This approach is capable of extension to other applications and hence the proposal here of developing a shopping list of options from which to choose.

An embryo shopping list is given in Table 4.6 for corrosion. Note that it is divided into three zones, A, B and C – broadly corresponding to material, design and construction matters, and there would probably have to be minimum requirements specified for each zone. Not all of the options are equal by any means, and there would be a need for development work to evaluate their relative merits. This is seen as a role for those involved in the performance prediction of concrete.

Ideally, each option should have a performance profile, perhaps defined by a numerical rating. A summation of ratings could then be matched against defined requirements for each zone in the required performance cube (Fig. 4.4) – say 10 units for zone A and 100 units for zone G.

The need for significant developments in this area is perhaps emphasized at present by the almost indiscriminate use of various protective measures, singly or in combination – either in new-build or in rehabilitation. The true performance of many of these, under real

Table 4.6 Shopping list of resistance options for corrosion

A Materials	Concrete quality	– mix proportions
		– mix ingredients
	Cover	– minimum
		– tolerances
	Permeable formwork	
	Concrete protection	– sealers
		– coatings
		– penetrants
		– layers
	Rebar protection	– epoxy coating
	Special steels	
	Non-corrodible reinforcement	
	Cathodic protection	
B Design	Design concept	
	Sructural detailing	
	Cladding, services, fittings, finishes	
	Articulation, joints, movement	
	Treatment of water, drainage	
	Control of the environment, barriers	
	Provision for inspection, maintenance, replacement	
	Accurate assessment of effects of deterioration mechanisms	
C Construction	Construction methods	
	Quality control	
	Certification	
	Testing	
	Rationalization, standardization, simplification	

conditions, is not well established. Indeed, in combination, some may even be incompatible – or produce side effects in other areas of performance. Although, at first sight, this proposal may be seen as akin to comparing apples and pears, the task becomes easier if the overriding objective is perceived as limiting corrosion, so that it does not affect safety or serviceability.

4.3.3 Loads and load effects

In designing for durability, a knowledge of the loads (and their effects) is more important than in structural design. How these effects are treated will depend on the level of calculation proposed in the design strategy.

The situation is summarized in Table 4.7. As a minimum, zone A will always be necessary, particularly if going for the prescriptive type of solution discussed in the introduction to this chapter. However, it is strongly recommended that, to be truly effective, the definition of exposure conditions is orientated towards specific deterioration mechanisms – put another way, the system in Table 4.2 is too vague; the approach should relate more to Table 4.1.

If the design stragegy takes us into zone B in Table 4.7, then it is important to estimate the local microclimate as accurately as possible, and, from that, predict the likely moisture state within the concrete itself. Predictive models exist, capable of doing that; however, the input is most important, and some appreciation of the nature of the moisture

Table 4.7 Durability loads and load effects – treatment in design

A	1. Identification of aggressive action by type and intensity 2. Definition of the outer environment by broad category (for level 1 'deemed-to-satisfy' design solutions)
B	3. Predictions of inner environment, where appropriate, and definition of most severe conditions for individual aggressive actions or relevant combinations of these 4. Modelling of deterioration processes, to provide service life calculations on a probabilistic basis, to meet agreed performance criteria (taken from the performance cube – Fig. 4.4). Modelling may have to be two-phase: (i) predicting deterioration itself, and (ii) predicting the effects of deterioration on resistance (bending, shear, bond)

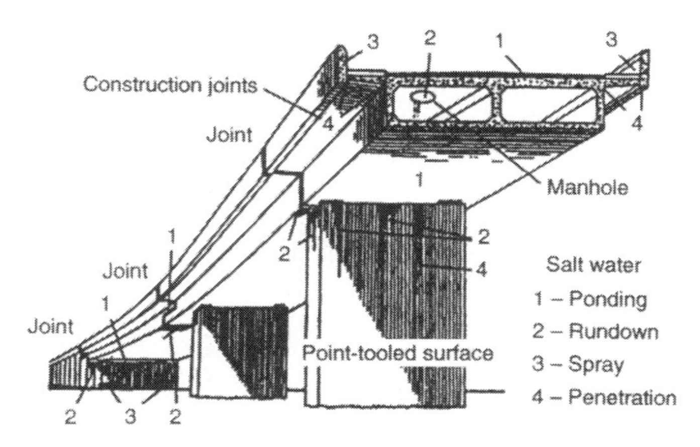

Figure 4.5 Various microclimate conditions for a bridge subjected to de-icing salts.

uptake is essential. This point has already been made in Fig. 4.2, and is reinforced in Fig. 4.5, this time for a bridge situation: corrosion rates will vary for the four different conditions.

Table 4.7 also stresses the need not only to predict deterioration itself, but also to assess the possible effects on structural resistance. Some types of structure are more sensitive to deterioration than others (e.g. post-tensioned concrete versus reinforced concrete) and the risk factor requires evaluation.

4.3.4 Risk and reliability design arena

This is illustrated in Fig. 4.6. The actions foreseen are shown in the diagram, and the notes indicate how the system would work. The current recipe approach (note 1(a)) will still be appropriate in some situations. However, if greater rigour/accuracy is considered necessary, then the essential elements are those in Table 4.4, and, for practical reasons, the approach has to be semi-probabilistic for design purposes. Risk analysis is suggested in support, really to explore the sensitivity of the structure to overload, i.e. what are the consequences if the design loads are exceeded?

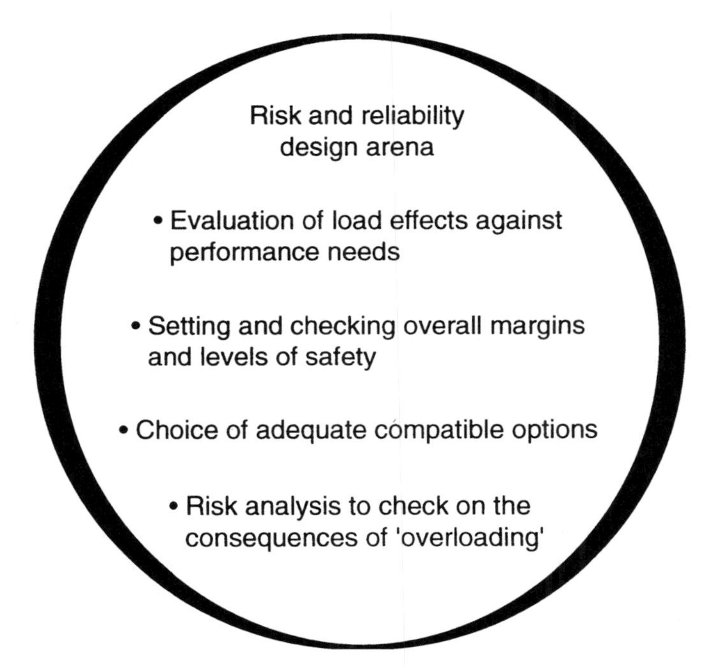

Figure 4.6 The risk and reliability design arena.

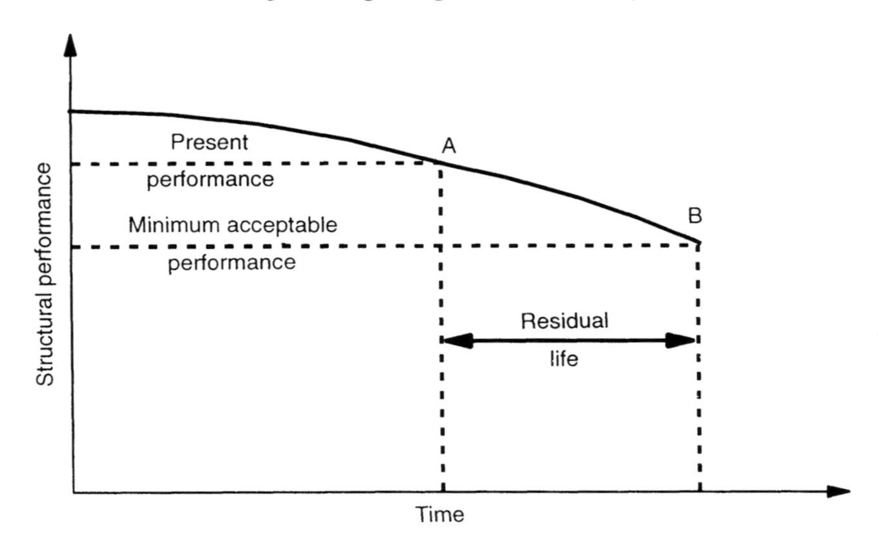

Figure 4.7 Schematic illustration of assumed behaviour in assessment work. (Reproduced from BCA, *The Residual Service Life of Reinforced Concrete Structures*; 1995.)

4.4 RELEVANCE OF THE APPROACH TO ASSESSMENT OF EXISTING STRUCTURES

It is considered that the approach outlined above is equally valid for assessment work; in fact, the basic principles were evolved as part of a major BRITE project on this subject [11]. The starting point may be different, as illustrated in Fig. 4.7, but the basic objective is the same – the determination of adequate strength and serviceability. The limiting criteria in Table 4.5 would come more into play, and there would be considerable emphasis on Table 4.7. By far the most difficult aspect would be the determination of minimum acceptable technical performance, but the BRITE project has demonstrated that this is feasible.

4.5 CONCLUDING REMARKS

In writing the paper ten years ago on design life [1], one of the author's selfish objectives was to clarify his own mind on the direction that future research on durability should take; many employees of the then Cement & Concrete Association were engaged on different aspects of the problem. The response to the paper [12] was somewhat disappointing in that the notional concept of a design life was taken literally, and there was discussion on liability when the notional period came to an end!

The concept has since gained some acceptance, but the author no longer believes that the specification of even a notional life is essential for durability design (Fig. 4.4). What **is** needed is a structured approach within a design framework, and it is to be hoped that this chapter gives some indication of how that might be done.

The major problem now is the same as it was ten years ago – all the essential pieces of the jigsaw are not yet in place. Knowledge has increased, but that knowledge is not in a form that can be used for routine design. In particular, the ability to predict deterioration has improved enormously, but some of the elements in Table 4.4 have not evolved in parallel. It is believed that this is now feasible, as the author's current BRITE experience has shown [11].

A significant development in the last ten years has been the growing awareness that provision for management, maintenance and replacement is essential. Nearly half the activity of the construction industry is in maintenance and refurbishment. Hopefully, this will provide the spur for the development (and application) of the framework described here, since the principles apply equally to new construction and assessment.

REFERENCES

1. Somerville, G. (1986) The design life of concrete structures. *The Structural Engineer*, **64A**, 60–71.
2. Wallbank, E.J. (1989) *Performance of Concrete in Bridges: A Survey of 200 Highway Bridges*. A report prepared for the Department of Transport by G. Maunsell & Partners, April, HMSO, London.
3. Anderson, J.M. and Gill, J.R. (1988) *Rainscreen Cladding: A Guide to Design Principles and Practice*. CIRIA Publication B5. Construction Industry Research and Information Association, London.
4. BS 8110 pt 1 (1985) *Structural Use of Concrete: Code of Practice for Design and Construction*, British Standards Institution, London.
5. Paterson, A.C. (1984) *The structural engineer in context*. Presidential address to the Institution of Structural Engineers, October 4, 1984. *The Structural Engineer*, **62A** (11), 335–42.
6. Concrete Bridge Development Group (1995) Whole Life Costing – concrete bridges. Proceedings of CBDG Seminar, April 25, CBDG, Crowthorne, UK.
7. White, K.H. (1991) Building performance and cost-in-use. *The Structural Engineer*, **69** (7), 148–51.
8. BS 7543 (1992) *Guide to Durability of Buildings and Building Elements, Products and Components*. British Standards Institution, London.
9. Somerville, G. (ed) (1992) *The Design Life of Structures*. Proceedings of the 1990 Henderson Colloquium, organized by the International Association for Bridges and Structural Engineering, July 16–18, 1990, University of Cambridge, Blackie, London.
10. Raiss, M.E. (1994) *Design Details for Durable Grouted Bonded Post-tensioned Concrete Bridges*. Proceedings of Concrete Society/Concrete Bridge Development Group Seminar, May 18, 1994. The Concrete Society, Slough, UK, pp. 73–106.

11. British Cement Association (1995) *The Residual Service Life of Reinforced Concrete Structures*. Proceedings of BRITE 4062 Workshop, April 6–7, 1995, Imperial College, London. BCA, Crowthorne, UK.
12. Somerville, G. (1986) The design life of concrete structures (discussion). *The Structural Engineer*, **64A** (9), 233–41.

5

Measuring and modelling transport phenomena in concrete for life prediction of structures

by N.R. Buenfeld

ABSTRACT

Most deterioration mechanisms affecting concrete structures are rate-controlled by the transport of particular aggressive species through the concrete. Consequently, an important part of predicting service life is generally to attempt to predict rates of ionic and molecular transport. This chapter examines some of the key issues concerned with predicting transport, taking chloride penetration as an example. The main transport processes affecting concrete structures are described. Some of the limitations of analytical solutions and models are exposed and the power of numerical modelling is demonstrated. Some of the key issues concerned with measuring transport coefficients are presented. Neural networks are introduced as a means of directly using data arising from condition surveys and exposure trials to predict transport.

Keywords: chloride, concrete, diffusion, durability, ionic and molecular transport, life prediction, models, neural networks.

5.1 INTRODUCTION

Inadequate durability is by far the most common cause of premature failure of concrete structures, yet very little attention is devoted to durability in the design process. Generally, durability is covered by prescriptive code recommendations based on previous code clauses, arbitrarily tightened where case histories have shown problems. However, this approach is suspect when applied to new materials or to design lives longer than relevant experience. Service lives of 120 years and longer are regularly being specified, even though reinforced concrete structures did not exist 100 years ago. Largely in recognition of these shortcomings, service life prediction is currently an area of significant research activity. There are essentially two areas of application of service life prediction; quantitative durability design of new structures and prediction of the residual service life of existing structures.

5.2 DEPENDENCE OF DETERIORATION MECHANISMS ON IONIC AND MOLECULAR TRANSPORT

The rates of most deterioration processes that affect concrete structures are controlled by the transport of particular aggressive species through the concrete, as outlined below.

- Reinforcement corrosion: the period until embedded reinforcement starts to corrode is generally controlled by the rate of penetration of chlorides or carbon dioxide into the concrete. The rate of corrosion is then dependent upon the availability and transport of water, oxygen and ions at the steel surface.
- Sulphate attack: the rate of sulphate attack depends on the rate of penetration of sulphate ions into the concrete.
- Frost attack: concrete has to be above a critical degree of water saturation for damage to occur during freezing. Rate of deterioration is therefore dependent upon rate of water penetration.
- Alkali–aggregate reaction: water is required to produce the expansive gel and it is therefore likely that the reaction rate will be dependent upon the rate of water penetration into the concrete. It is also likely that the transport of alkalis and gel will affect the rate of deterioration due to alkali–aggregate reaction.
- Abrasion: this is the only common deterioration process affecting concrete structures that is not directly controlled by ionic or molecular transport.

5.3 TRANSPORT PROCESSES

There are a number of different transport processes that may participate in the deterioration of a concrete structure. These are listed below, together with the transport coefficients characterizing the process.

- Pressure-induced water flow: flow of water due to the application of an hydrostatic head. Characterized by a water permeability coefficient, this transport process has been researched more thoroughly than any other. However, it is rare for this to be the predominant transport process; exceptions include water-retaining structures and deeply submerged concrete.

- Water absorption: uptake of water resulting from capillary forces, characterized by a sorptivity coefficient. In environments where significant concrete drying is possible, water absorption may lead to very rapid penetration of species dissolved in the water. For example, concrete in the tidal zone of an offshore structure, or bridge decks subjected to regular applications of de-icing salts, may suffer from rapid chloride ion ingress due to water absorption.
- Water vapour diffusion: diffusion of water as a vapour, characterized by a water vapour diffusivity coefficient. This is the process by which concrete dries and it controls the moisture distribution in a concrete element after a period of drying. In turn, moisture distribution influences properties such as water absorption during wet and dry cycles, electrical resistivity, gas permeability and gas diffusivity.
- Wick action: a combination of water absorption and water vapour diffusion, characterized by **sorptivity** and **water vapour diffusivity** coefficients respectively. Wick action is the transport of water from the wetted face of a concrete element to a drying face. Species dissolved in the water are transported to the zone of the element where drying occurs and may result in salt crystallization.
- Ion diffusion: movement of ions as a result of a concentration gradient, characterized by an ion diffusion coefficient. Ion diffusion is only significant where the concrete is nearly or completely water saturated. The ions that are usually of most interest are chlorides and sulphates. Compared with most other transport processes, ion diffusion is very slow, but is often rapid enough to cause deterioration well within the design life of a concrete structure.
- Gas diffusion: movement of gas molecules as a result of a concentration gradient, characterized by a gas diffusion coefficient. Diffusion of carbon dioxide into concrete results in carbonation. In some situations, diffusion of oxygen controls the rate of corrosion of steel in concrete.
- Pressure-induced gas flow: flow of gas due to a pressure gradient, characterized by a gas permeability coefficient. Gas permeability is relatively easy to measure, but is not relevant to the behaviour of the vast majority of concrete structures. It is relevant to the behaviour of certain structures used in the nuclear industry.

Maximum values of transport coefficients are now being included in some concrete specifications and this is likely to become more prevalent. The challenge is to develop methods of assessing compliance that can be undertaken sufficiently rapidly to be appropriate within a construction contract. Methods of measuring transport coefficients are discussed in section 5.5.

5.4 MODELLING TRANSPORT

5.4.1 General

For service life prediction to be possible, the progress of deterioration mechanisms likely to be contributing to loss of serviceability must be quantified, i.e. the deterioration mechanisms must be modelled. The model is the set of equations and boundary conditions used to represent the mechanism. Because of the dependence of most deterioration mechanisms on transport, most models of deterioration mechanisms are essentially models of transport processes. Service life (or a part of it, see discussion of initiation and propagation times below) is generally inversely related to the relevant transport coefficient(s) and potential(s). Transport coefficients are properties of the concrete, as defined in section 5.3 and key issues relating to their measurement are presented in section 5.5. The potential is a function of the exposure environment; it can be viewed as the driving force that causes transport. For pressure-induced water or gas flow the potential is the applied pressure. For ion and gas diffusion it is the concentration differential. Water vapour diffusion is a special case of gas diffusion and the potential can be expressed as either a concentration differential or a relative humidity differential. Sorptivity is a convenient method of characterizing water absorption and does not require the definition of a potential to make predictions.

If the model is simple, it may be possible to obtain an analytical solution, involving simple substitution of values into an equation to make a prediction. If an analytical solution is not possible, a numerical solution is required, generally an iterative computer-based approach. The term modelling is used to describe selection of the model, selection and development of a solution, and application of the solution to make predictions.

Generally the model and solution are selected and then the corresponding transport coefficients, potentials and other parameters required are determined. Ideally solutions should involve only known or measurable variables.

Chloride-induced reinforcement corrosion is the most widespread and costly deterioration mechanism affecting concrete structures and it is therefore this mechanism that is selected to illustrate the different aspects of modelling. The service life of a reinforced concrete structure in a chloride-rich environment can be considered to consist of two periods; an initiation period (t_0) prior to the chloride concentration at the steel reaching the critical level required for corrosion, and a propagation period (t_1) during which corrosion occurs, until an unacceptable amount of damage occurs. Initiation and propagation involve quite different processes and it is therefore usual to model them separately. In this chapter models and solutions of the initiation period are presented. These could be arranged so that t_0 is predicted directly. However, it is more informative to predict chloride profiles, delaying the choice of cover depth and critical chloride level.

5.4.2 Analytical solutions

Fick's law, describing diffusion of an unreactive species held at a fixed concentration into a semi-infinite medium, is conventionally selected as the model for chloride penetration into concrete in real environments. The analytical solution is an error function equation as follows.

$$C_x - C_b = (C_s - C_b)\left[1 - \text{erf}\frac{x}{2(Dt)^{1/2}}\right], \quad (1)$$

where:

C_x = chloride content at depth x
C_s = chloride content at surface
C_b = background (i.e. from the mix ingredients) chloride content
D = effective chloride diffusion coefficent
t = exposure period.

The most common application of this equation is to extrapolate from the chloride profile (i.e. C_x vs. x) measured after a relatively short period of exposure, to predict t_0. In the case of residual service life prediction, chloride profiles measured during a condition survey would be used (measured profile in Fig. 5.1). In the case of a new structure, chloride profiles measured in specimens of similar concrete after a period of immersion in a chloride solution would be used.

In theory, D is the only unknown in (1) and hence it can be calculated by measuring C_s, C_b and C_x at a single value of x. However, direct measurement of C_s is unreliable as C_s is the chloride content in the concrete right at the exposed surface and if the depth increment were small enough to represent the surface it would probably not be representative of the concrete. Furthermore, C_s may vary with time, for example it may reduce if the surface is washed by rain. The usual approach, therefore, is to fit a curve to the chloride profile with C_s and D as the independent variables (fitted profile in Fig. 5.1).

In the case of residual service life prediction, these values of C_s and D can then be used to predict the time (t_0) when the chloride content at cover depth x reaches the critical chloride level, generally taken as 0.4% chloride by weight of cement, as shown by the predicted profile in Fig. 5.1. In the case of a new structure the calculated value of D would be used, but if the concrete were exposed to a different concentration

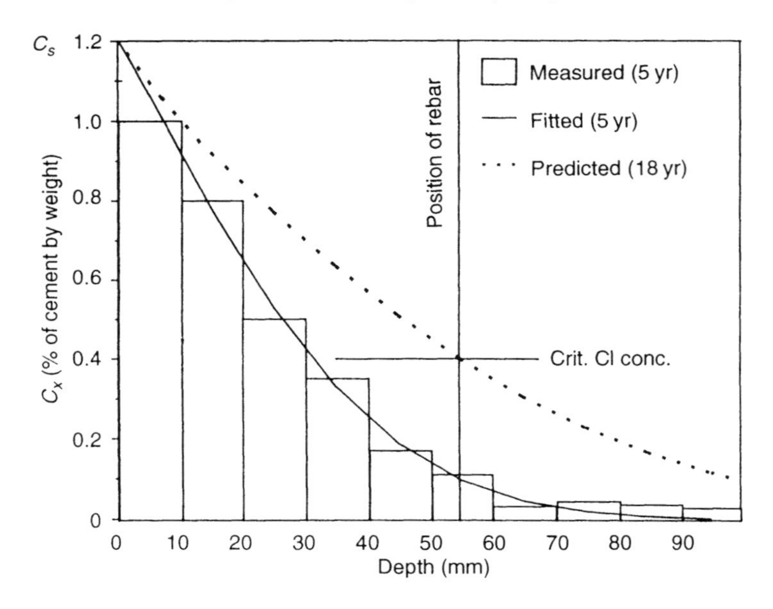

Figure 5.1 Measured, fitted and predicted chloride profiles.

solution in the test (generally a higher concentration to accelerate the short-term test) than in practice, a more appropriate value of C_s should be chosen.

It should be appreciated that, as described, this approach does not involve the application of a factor of safety; research is currently being directed at determining the most appropriate method of doing this.

Unfortunately this conventional approach of predicting chloride profiles is limited, and often flawed, for a number of reasons. Firstly chloride is reactive; a proportion of chlorides entering concrete is chemically bound by cement paste constituents with only a proportion remaining free to diffuse. Furthermore, chloride binding is probably non-linear with most of the chloride present bound at low chloride contents, with the bound proportion decreasing with increasing chloride content. Secondly, D generally reduces with time as the cement continues to hydrate; this is particularly significant for concretes containing PFA. Thirdly, in most situations where chloride-induced corrosion is a problem, ion diffusion is unlikely to be the only transport process responsible for chloride penetration; transport processes such as water absorption (during wetting and drying cycles), pressure-induced flow of water and wick action may also be involved. There is no reason why Fick's law should apply to these other processes. These drawbacks can be overcome by numerical modelling.

5.4.3 Numerical solutions

Numerical models generally involve dividing the concrete into a large number of discrete elements; in uniaxial penetration problems this is generally a series of laminae, each parallel to the exposed concrete surface. Initial boundary conditions are set on each side of the concrete and physical properties (e.g. transport coefficients) and possibly chemical properties (e.g. a chloride binding isotherm) are attributed to each element. The equations governing behaviour are then solved for each element for a small step forward in time; the resulting values are used in the next time increment.

Here the inputs and corresponding output (Fig. 5.2) from a finite-difference model of chloride transport into concrete immersed in sea water are presented to demonstrate how the numerical approach overcomes the drawbacks highlighted in section 5.4.2.

Common inputs for all of the predicted profiles presented in Fig. 5.2 are:

- sea water chloride concentration: 20 g/1
- concrete density: 2400 kg/m^3
- cement content: 400 kg/m^3
- accessible porosity: 12%
- background chloride content: 0%
- water permeability: 10^{-13} m/s
- element thickness: 200 mm
- exposure period: 50 years

The other inputs required, which were varied to produce the profiles in Fig. 5.2, are presented in Table 5.1.

Ion diffusion is again modelled by Fick's law. Profile 1 in Fig. 5.2 is the predicted profile after shallow sea water immersion, i.e. the hydrostatic head is negligible so that chloride transport is due to ion diffusion alone. Profile 1 does not involve chloride binding (this is a hypothetical case as some degree of binding always occurs) and so all of the chloride is free (i.e. unbound). Transport and chloride binding are treated separately and their effects then combined. Chloride binding is modelled by a binding isotherm which determines the proportion of the total chloride present that is bound. The binding isotherm is dependent upon the cement type (including mineral additions), concrete mix proportions and certain environmental factors. There are currently insufficient data in the literature to allow selection of a binding isotherm specific to a

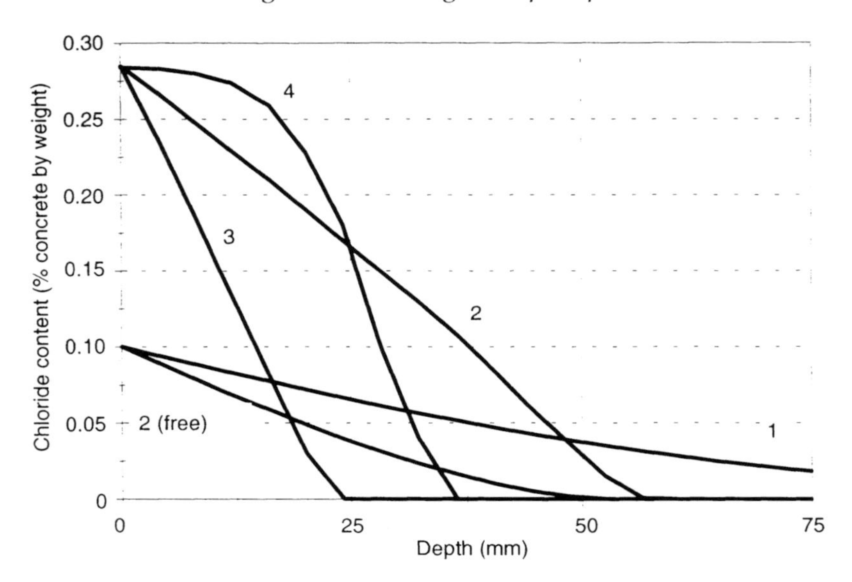

Figure 5.2 Chloride profiles predicted by finite-difference model (total chloride profiles except where indicated).

Table 5.1 Input variables for chloride profiles presented in Fig. 5.2

Profile	*D (m^2/s)*	*Binding*	*Head (m)*
1	10^{-12}	×	0
2	10^{-12}	✓	0
3	10^{-12} to 10^{-13} over first five years	✓	0
4	10^{-12} to 10^{-13} over first five years	✓	10

particular concrete and set of exposure conditions. Figure 5.3 presents data points drawn from the literature, covering a range of concretes.

The curve shown in Fig. 5.3 is a composite of a Langmuir isotherm at low concentrations and a Freundlich isotherm at higher levels. There is a physico-chemical basis for choosing such isotherms and it can be seen that the composite isotherm fits the data reasonably well. Profile 2 has identical input variables to profile 1 except that chloride binding occurs. It can be seen that binding dramatically reduces the depth of penetration, but results in higher chloride contents in the penetrated zone. The free chloride component of profile 2 is also indicated in Fig. 5.2.

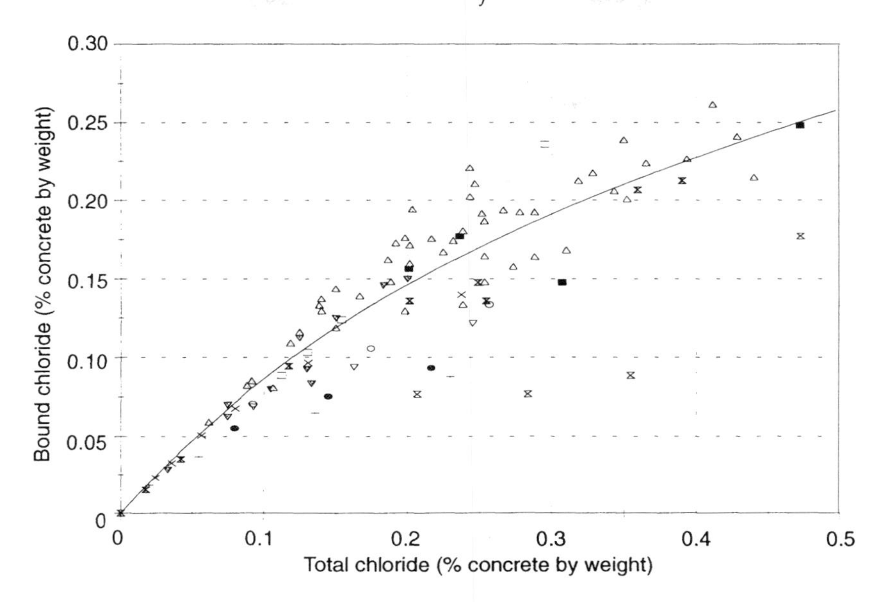

Figure 5.3 Bound vs. total chloride data.

Because the finite-difference method involves stepping forward in time in small increments, properties of the concrete may be varied with time. For example, profile 3 is the result of an order of magnitude linear reduction in D over the first five years of sea water immersion, all other inputs being the same as for profile 2. Additionally, concrete properties may be varied with depth, to take into account the effect of curing, environmental exposure, etc.

As the depth of immersion increases, the contribution of pressure-induced flow of sea water into the concrete becomes significant. Pressure-induced flow is modelled by Darcy's equation. If it is assumed that pressure-induced water flow and ion diffusion occur through the same porosity, their effects are easily combined. Profile 4 is the result of 10 m head of sea water, all other inputs being the same as for profile 3.

The influence of temperature gradients and cycles could be taken into account by defining the temperature dependence of the transport coefficients involved. The numerical approach also allows modelling in more than one dimension and inclusion of complex boundary conditions.

Numerical modelling is a powerful approach and, in the author's view, is likely to become more prevalent as the field of service life prediction develops. However, as numerical models become more sophisticated, more inputs are involved, generally requiring improved characterization of exposure environments and determination of more concrete properties than are required for analytical solutions.

5.5 MEASURING TRANSPORT COEFFICIENTS

Most models of the type described in section 5.4 require the input of one or more transport coefficients. In the case of residual service life prediction it may be that transport coefficients can be calculated based on measurements of the progress of deterioration of the structure in question; for example chloride profiles or carbonation depths. In the case of life prediction of new structures, transport coefficients pertaining to the concrete in question may be available in the literature, or it may be necessary to measure them.

Worldwide, there are few transport test methods that have been accepted as national standards. The only British Standard transport test method is the initial surface absorption test (ISAT) [1]. This was primarily developed for *in situ* measurements and is of limited value in life prediction because of the lack of control of moisture content of the concrete and the fact that the volume of concrete influencing the test result is poorly defined. However, there are a number of laboratory methods available for measuring each of the transport processes listed earlier. There is not the space to run through the various test methods in this chapter; Concrete Society Technical Report 31 [2] is a good source of information. The remainder of this section addresses some of the key issues faced when selecting or devising a transport test method.

Where possible the transport coefficient of interest should be measured directly. However, occasionally it is preferable to measure a different transport coefficient that can be correlated to the coefficient of interest.

In most transport tests, the transport process of interest is accelerated in relation to its rate in practice, in order to reduce the test duration. In a water-permeability test a particularly high hydrostatic head may be used. In an ion-diffusion test the process may be accelerated by raising the temperature or by applying an electric field [3]. The more a transport process is accelerated, the more it is likely to deviate from the process that occurs naturally. Tests should therefore be accelerated only as much as is necessary. Clearly, testing for compliance during production will require a very rapid test method. However, pre-production mix development trials should be started as early as possible to allow less accelerated methods to be used.

Using tests that measure a single transport process, to determine a well-defined transport coefficient, requires conditioning specimens to a known moisture state prior to testing. In the case of water permeability and ion diffusion, specimens should be water saturated. In the case of water absorption, gas diffusion and gas permeability specimens should be pre-dried. It should be remembered that conditioning time will generally be inversely related to the penetrability of the concrete. The main drawback of site-based test methods (such as the ISAT, CLAM and Figg methods) is that the concrete being tested cannot be properly preconditioned. Under certain circumstances site measurements are useful as qualitative indicators of concrete penetrability, but in the context of life prediction, it is generally preferable to take a core and carry out transport testing in the laboratory.

Most transport tests involve measuring transport between two opposite, flat faces of a specimen, with the circumference of the specimen sealed into the test apparatus. Cylindrical discs are therefore a convenient shape; discs may be cast to size, or cylinders or cores may be sliced. The thickness should be such that interfacial zones, between aggregate particles and cement paste, do not create a short-circuit path through the specimen; a thickness of 2.5 times the maximum aggregate size is usually sufficient (100 mm diameter by 50 mm thick is the usual size of specimens tested at Imperial College). Consideration should also be given to whether bulk or surface concrete should be tested. If the study is associated with the protective properties of the concrete cover (e.g. keeping chlorides from the reinforcing steel), then the concrete surface should be tested. If the study is concerned with the properties of the full thickness of a concrete element (e.g. water transmission through a basement wall), then it is probably best to test a specimen from the bulk of the concrete, excluding the concrete surface. Other issues to address include:

- how many replicates to test;
- the possibility of testing each specimen for a range of properties, if the later tests are not influenced by the earlier testing regime;
- the possible presence of contaminants affecting the validity of the test method;
- the amount of similar test data with which the results may be compared.

5.6 NEURAL NETWORKS

In some situations modelling of a deterioration process may be very difficult, or it may be impractical to measure all of the transport coefficients and other controlling parameters required. Some transport processes are so slow, or the transport coefficients change significantly over such long periods of time, that laboratory measurements are unreliable. There are also some aspects of real construction that may affect durability that are difficult to reproduce in the laboratory such as slip-forming and heat of hydration effects in thick elements. In these situations the approaches to service life prediction outlined in sections 5.3 to 5.5 may not be practical.

Much data are arising from condition surveys of concrete structures and from long-term natural exposure trials on concrete specimens. Ideally these would be used as the basis for predicting service life. Unfortunately, this information generally involves a large number of relevant variables, missing data, non-linearity and other features rendering it very difficult to use with conventional methods of data analysis. Research undertaken at Imperial College has shown that neural networks (NN) offer a way forward [4–6].

Neural networks are computing devices based on simple adaptive models of neurons (nerve cells). An NN is formed from tens or perhaps hundreds of simulated neurons connected to each other, largely in parallel, in much the same way that neurons are connected in the brain. As an example, Fig. 5.4 shows the structure of a simple NN to predict chloride profiles in concrete exposed to a marine environment. This NN

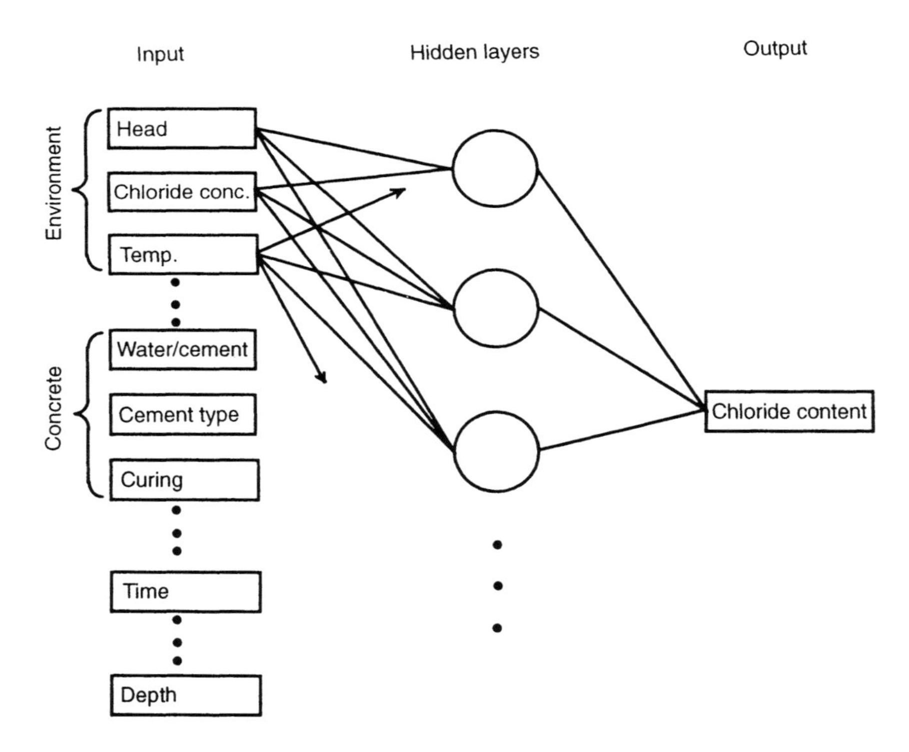

Figure 5.4 Neural network for predicting chloride profiles.

has three layers of neurons: an input layer, a hidden layer and an output layer. Each variable that is thought to possibly influence chloride penetration is represented by an input neuron. Therefore there are inputs relating to environmental factors such as exposure zone (tidal, splash, etc.), material factors such as water/cement ratio, time and depth. The output is the chloride content.

Each neuron receives the output signal from many other neurons. A neuron calculates its own output by finding the weighted sum of its inputs, generating an activation level and passing it through a transfer function. The point where two neurons communicate is called a 'connection'. The strength of the connection between two neurons is called a 'weight'. The most popular learning method is by example and repetition; the NN is presented with a set of inputs together with the correct corresponding outputs; each time an input is presented the network sends back an answer of what it thinks the output should be. This is compared with the correct output. If the output is incorrect, the network makes corrections to the weights on the connections between neurons. This training process is repeated until the NN outputs are correct. The NN is then tested to determine its accuracy by presenting it with sets of input data (not previously seen by the NN) and comparing the outputs with the known correct outputs. Finally, the NN is used, i.e. it is presented with real input data and predicts unknown output values. A chloride profile is predicted by running the NN for a range of depths.

NN are good at handling a large amount of data. However, an NN is only as good as the data on which it is trained. A moderate amount of training data spread across the whole domain of interest is more valuable than a vast amount of data limited to only a subset of all of the variables of interest.

The NN approach requires knowledge of the variables that may be important, but does not require prior understanding of the relationships between each variable and the output of interest. An NN predicts future chloride profiles directly, without introducing errors by going through the intermediate step of calculating a D and C_s value. Predictions beyond the time range of the training data are achieved by using the NN to determine the time dependence, but then requires fitting of a function to allow extrapolation to longer times.

An NN may be a practical method of encompassing results from a large number of different studies to validate or modify design code durability recommendations.

ACKNOWLEDGEMENTS

The author would like to thank Ian McLoughlin (research assistant) for his assistance in numerical modelling and preparation of figures.

REFERENCES

1. British Standards Institution (1970) *Testing Concrete: Methods of Testing Hardened Concrete for Other than Strength*, BS 1881 pt 5, BSI, London.
2. Concrete Society (1987) *Permeability Testing of Site Concrete – A Review of Methods and Experience*. Tech. Rep. 31; Concrete Society, Slough, UK.
3. Zhang, J.-Z. and Buenfeld, N.R. (1994) Development of the accelerated chloride ion diffusion (ACID) test, in *Proceedings of the International Conference on Corrosion and Corrosion Protection of Steel in Concrete*, July 24–29, 1994, University of Sheffield. Sheffield Academic Press, UK, pp. 395–403.
4. Buenfeld, N.R. and Hassanein, N.M. (1995) Neural networks for predicting the deterioration of concrete structures, *NATO/RILEM Workshop on the Modelling of Microstructure and its Potential for Studying Transport Properties and Durability*, St Remy-les-Chevreuse, France, July 10–13, 1994, Kluwer Academic Publishers, pp. 415–432.
5. Buenfeld, N.R. and Hassanein, N.M. (1996) An artificial neural network for predicting carbonation depth in concrete structures, in *2nd ASCE Monograph on Artificial Neural Networks in Civil Engineering*. American Society of Civil Engineers.
6. Glass, G.K., Hassanein, N.M. and Buenfeld, N.R. (1997) Neural network modelling of chloride binding, in *Magazine of Concrete Research*, Thomas Telford.

6

Aggregates: a review of prediction and performance

by P.G. Fookes

A survey . . . of the common British aggregates *has failed to reveal any containing alkali–reactive constituents,* but care is needed when using aggregates of geological types which might contain such constituents and of which no previous service experience is available. [!!] (author's italics)

Source: Lea, F.M. The Chemistry of Cement and Concrete, 3rd edn; published by Edward Arnold, 1970.

ABSTRACT

This chapter starts with an overview of aggregates in concrete, deterioration mechanisms and an evaluation of deterioration related to aggregates. It then briefly looks at rock material and rock masses, in order to give background and to put various geological materials and terminology in context for the following longer discussions on identification and on some unsatisfactory physical and chemical characteristics of aggregates. It concludes that no rigorous models yet exist for predicting the deleterious performance of aggregates and there is no substitute for testing potential aggregates in advance.

Keywords: aggregate–mineral reaction, alkali–aggregate reaction (AAR), alkali–silica reaction (ASR) concrete investigation, deleterious reaction, durability, geology, petrography, aggregate testing.

6.1 INTRODUCTION

There are a number of potential defects in natural aggregates. These often relate to the presence of minor constituents. In some uses these defects may be of little or no importance but in others they may be at least aesthetically displeasing and at worst limit the useful life of the structure. Such problems can be avoided by early and suitable investigation of the materials to be employed. Steps taken to prevent deterioration may not be seen to be effective for many years so that it is prudent to err strongly on the side of caution.

Source: French, W.J. [1]

6.1.1 Background

The main purpose of this review is to bring together and discuss properties of aggregates and their potential for affecting the aggregates' performance in concrete. It is intended mainly for natural concrete aggregates in the UK but, where appropriate, common problems with aggregates overseas are mentioned where the local environment modifies the ways in which the problems occur or develop.

The chapter starts with an overview of aggregates in concrete, deterioration mechanisms and the evaluation of deterioration related to aggregates. It then briefly looks at rock material and rock masses, in order to give background and to put various geological materials and terminology in context for the following longer discussions on identification and on some unsatisfactory physical and chemical characteristics of various aggregates.

Contributions that geologists can make to the study of aggregates stem from an understanding that rock owes its properties to its origin, its mineral composition and the geological processes that have affected it through time. This helps them to determine the suitability of rock for use as aggregates and to make an informed search for new deposits. It also helps them to recognize and define the generally clear but not always simple relationships that exist between the rock – its mineral composition, texture, grain size, fabric, state of weathering, alteration and contamination, and its likely performance as an aggregate when used in concrete.

An all-embracing preview of natural concrete aggregates would be very extensive and for many aggregates would be complex depending on the way in which they react with cement, additives or water, and how they perform within the concrete. This review does not attempt to study these problems in detail, but does note many of them. The report by the Geological Society Engineering Group Working Party on

Aggregates [2] and Fookes [3] give more detail about aggregates and the references they list can lead to wider reading.

'Soundness' is used throughout this review generally to mean the ability of an aggregate to resist excessive volume changes as a result of changes in the physical environment, e.g. freeze–thaw, thermal changes at temperatures above freezing, slaking (alternate wetting and drying), i.e. soundness is commonly used to indicate physical and mechanical soundness. The rocks most susceptible to such volume changes are dolerite, porous chert, shales and any rocks with a significant content of clay minerals, e.g. montmorillonite or illite. Used in concrete, such material could lead to deterioration through scaling or more extensive cracking.

The soundness of a rock may be decreased through natural weathering or alteration. Natural weathering, i.e. producing 'weathered rock', occurs on a geological time scale, but in some situations it can occur in service over a period of months or years [4]. Slight to moderate geological weathering is typically manifest by an increase in porosity and water absorption and this in turn can affect the response of the rock in service to its environment should such rock be used to make aggregates.

Chemical deterioration of aggregates in service is primarily due to chemical reactions with components within the concrete, although on occasion chemical reactions may be brought about by external forces within the service environment of the concrete. All such reactions are considered as chemical reactions and can occur to a wide range of rock types and are discussed in context in section 6.4. Such reactions may loosely be called 'chemical unsoundness'.

Extensive use is made of tables and figures throughout to summarize and synthesize some of the complexities of the geological jargon; in addition, photomicrographs of aggregates in concrete are used to illustrate some of the problems.

6.1.2 Aggregates in mix design

It is not the prime purpose of this review to look at the physical and mechanical properties of aggregates. In the UK technical literature there are several excellent publications on the subject, e.g. Building Research Establishment RR18 [5]; BRE Information Papers IP 2/86 [6], IP 16/89 [7], IP 7/87 [8]; RMC [9], C&CA [10], BCA [11], Property Services Agency [12], BS 4721 [13] and BS 1199 [14]. The BRE Digest 237 [15] is recommended as the most appropriate reading on the subject, and most text books seem to have drawn heavily from this source. The BRE Digest, in turn, draws from the work of Teychenné [16].

Aggregates make up the major part of concrete and their properties

Table 6.1 Properties of concrete using different aggregates[a]

Aggregate	*Typical range of dry density*		*Compressive strength (N/mm²)*	*Drying shrinkage (%)*	*Thermal conductivity at 5% moisture content[b] (W/mK)*	*Main functional requirement*
	Aggregate (kg/m³)	*Concrete (kg/m³)*				
Flint gravel or crushed rock	1350–1600	2200–2500	20–180	0.03–0.08	1.6–2.2	Class 2 fire resistance; Strength and durability
Crushed limestone	1350–1600	2200–2450	20–150	0.03–0.04	1.6–2.0	Class 2 fire resistance; Strength and durability
Crushed brick	1100–1350	1700–2150	15–30	–	0.85–1.50	Class 2 fire resistance; Class 1 fire resistance; Strength and durability
Expanded clay, shale or slate and sintered pulverized fuel ash	300–1050	1350–1800	15–60	0.02–0.12	0.55–0.95	Class 1 fire resistance; Thermal insulation; Strength and durability
Foamed slag	500–950	1700–2100	15–60	0.04–0.10	0.85–1.40	Class 1 fire resistance; Thermal insulation; Strength and durability
Expanded clay, shale or slate and sintered pulverized fuel ash	300–1050	700–1300	2–7	0.03–0.07	0.24–0.50	Class 1 fire resistance; Thermal insulation; Strength and durability
Foamed slag	500–950	950–1500	2–7	0.03–0.07	0.30–0.50	Class 1 fire resistance; Thermal insulation
Pumice	500–900	650–1450	2–15	0.04–0.08	0.21–0.63	Class 1 fire resistance; Thermal insulation
Exfoliated vermiculite and expanded perlite	60–250	400–1100	0.5–7	0.20–0.35	0.15–0.39	Class 1 fire resistance; Thermal insulation
Clinker	700–1050	1050–1500	2–7	0.04–0.08	0.35–0.65	Class 1 fire resistance; Thermal insulation

[a] Source: BRE Digest 237, *Materials for Concrete*; published by HMSO, 1980.
[b] Calculated from Mitchell [17]; actual measured conductivities should be used in preference to these for calculating standard U-values.

affect the physical characteristics of the concrete, such as density, strength, thermal conductivity and elastic thermal shrinkage (Table 6.1) and creep movements. Table 6.1, modified after BRE Digest 237 [15], gives data on both natural and synthetic aggregates for comparison.

The shape and surface texture of the aggregate particles and their grading, i.e. the distribution of particle sizes, are perhaps the most important factors influencing the workability and strength of the concrete. Teychenné [16] describes an investigation into the characteristics of concrete made with coarse and fine crushed-rock aggregates, samples of which were obtained from some 24 quarries, covering a wide variety of materials and geographical distribution throughout the UK. His main programme was concerned with the crushing strength and other mechanical properties of concrete and involved casting some 550 concrete mixes, together with a further 52 mixes to examine other characteristics of concrete associated with durability. Figure 6.1 summarizes some of his work in a telling fashion by showing the water requirements of 24 mixes using different aggregates to produce concretes of the same workability.

In essence, Teychenné concluded that good quality concrete could be made with all the aggregates he used; mechanical and durability problems of the concrete varied depending on the aggregate and the differences in the performance of the concrete could not be related to any single characteristic of the aggregate. Notably, his work showed that materials often regarded as of inferior quality for use in concrete

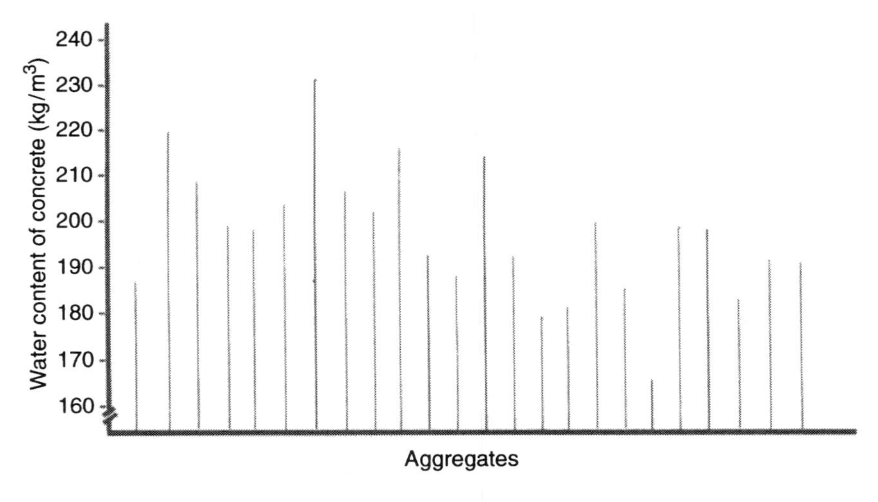

Figure 6.1 Water requirements of 24 mixes using different aggregates to produce concrete of the same workability. (Redrawn from Teychenné, *The Use of Crushed Rock Aggregates in Concrete*; BRE, 1978.)

may be quite suitable for that purpose. For example, he found that increasing the dust content of the fines from 10% to 25% resulted in only a small decrease in the crushing strength of the concrete, and the concrete of a required crushing strength could be obtained with the aggregates used by adjustment of the mix proportions; other characteristics such as elastic modulus or drying shrinkage depended primarily on the aggregate and were less affected by changes in the mix proportion.

Natural aggregates, such as sands, gravels and crushed rock, should comply with BS 882 [18] or BS 1201 [19]. The main functional requirement of concrete made with these aggregates is to provide strength and durability. The concrete may also be required to provide fire protection, which would depend on the aggregate used.

The grading of the fine aggregate (i.e. material passing a 5 mm sieve) has a particularly marked influence on the properties of the concrete. The grading of the fine aggregate should comply with one of the three grading zones specified in BS 882 [18] although gradings falling outside these zones may indeed still provide good aggregates, subject to trials.

Aggregates are commonly specified to be not of a certain type nor to contain impurities that react adversely with cement or embedded materials. The use of aggregates that have a high drying/shrinkage is also commonly banned, details being given in BRE Digest 35 [20] on shrinkage of natural aggregates in concrete.

6.2 AGGREGATES AND DURABILITY

6.2.1 Deterioration mechanisms and progress

If a concrete has been designed and well constructed to suit the performance required of its service life, apart from considerations of excessive use producing more than fair wear and tear, or the structure becoming uneconomic to run, or growing out of date, the life of the concrete will be determined by its rate and mode of deterioration. This is a large subject and publications on deterioration are currently a growth area of which the deterioration and performance of aggregates are only a relatively small part. A good practical document is the recent CEB design guide for durable concrete structures [21].

Figure 6.2, modified from the CEB publication [21], shows the CEB

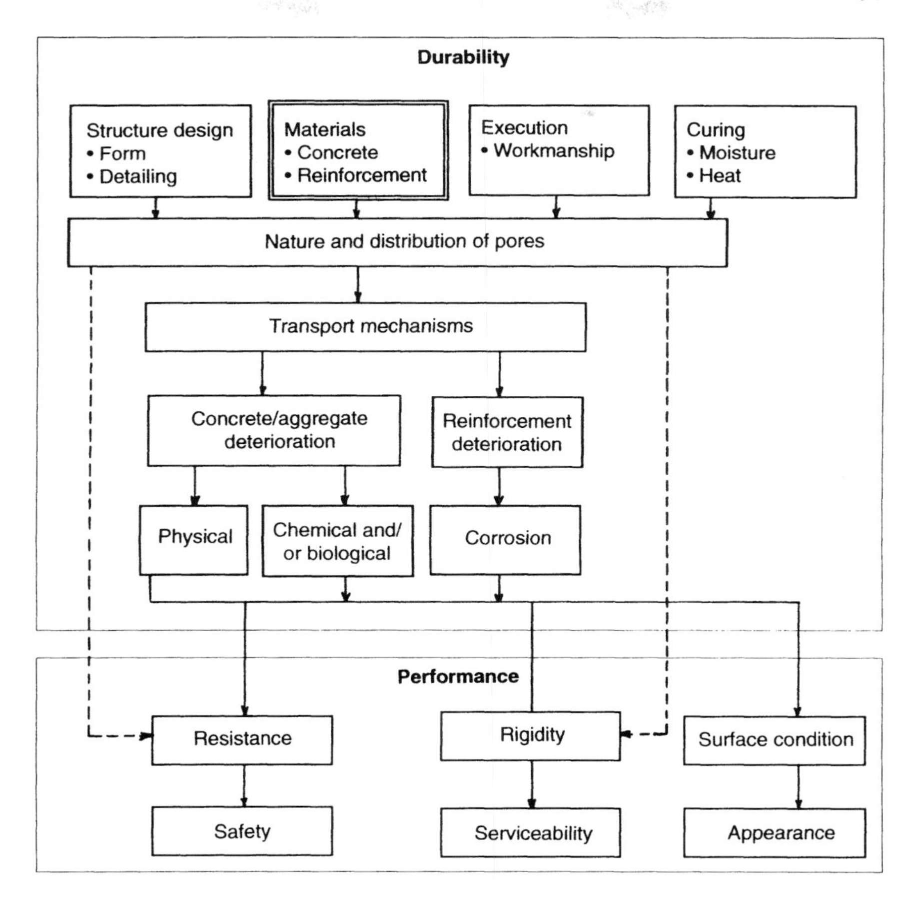

Figure 6.2 Relationship between the concepts of concrete durability and performance. (Redrawn from Comité Euro-International du Bréton, *Durable Concrete Structures: Design Guide*, 2nd edn; Thomas Telford, 1992.)

view of the relationship between the concepts of concrete durability and performance.

As can be seen, materials form one of the four prime inputs, together with structural design, execution (workmanship) and curing, which lead to the performance of the concrete. In describing this table, the CEB design guide says:

> It can be seen that the combined transportation of heat, moisture and chemicals, both within the concrete mass and in exchange with the surroundings (the microclimate), and the parameters controlling these transport mechanisms, constitute the principal elements of durability. The presence of water or moisture is the

single most important factor controlling the various deterioration processes, apart from mechanical deterioration. The transport of water within the concrete is determined by the pore type, size and distribution and by cracks (microcracks and macrocracks). Thus, controlling the nature and distribution of pores and cracks is essential. In turn, the type and rate of degradation processes for concrete (physical, chemical and biological) and for reinforcing or prestressing reinforcement (corrosion) determine the resistance and the rigidity of the materials, the sections and the elements making up a structure.

Table 6.2 Main parameters influencing the deterioration of concrete structures

Deterioration in reinforced concrete

The four most important mechanisms:

- Reinforcement corrosion (e.g. chlorides, carbonation)
- Aggregate unsoundness (e.g. alkali–aggregate reactions, AAR, ASR, ACR); aggregate mineral reaction, AMR
- Chemical attacks (e.g. sulphate, acids)
- Freeze–thaw scaling (e.g. frost)

Corrosion primarily destroys the reinforcement and the three others primarily destroy the concrete

	Prime deterioration mechanism			
Influencing parameter	*Corrosion*	*Aggregate unsoundness*	*Sulphate attack*	*Frost attack*
Environmental	x	x	x	x
Exposure	x	x		x
Cover	x			
Cement type	x	x	x	
Porosity	x	x	x	x
Air void characteristics				x
Cracks	x			x
Delamination	x			
Electrochemical potential of reinforcement	x			
Degree of carbonation	x			
Chloride content	x			
Degree of moisture saturation	x	x	x	x
Residual alkali–silica reactivity	x			
Sulphate content		x		

In this can be seen a common view, which is also the view of the author, that by implication, if good aggregates with sound properties and appropriate physical and mechanical characteristics are selected then, by and large, deterioration is not directly related to the aggregate. This is not to diminish the part played by the aggregate, but merely to highlight the reasonable presumption in countries with well-established aggregate practice, that good aggregates with sound characteristics and appropriately selected physical and mechanical characteristics go on performing quietly and effectively in concrete year after year. The selection of inappropriate aggregate, which sometimes produces spectacular deterioration (e.g. alkali–silica reaction, ASR), is fortunately relatively minor compared with the large amount of concrete produced. Nevertheless, it is also important to say that the part played by physically or chemically unsound aggregates must not be overlooked or underestimated.

Table 6.2 very broadly summarizes in a somewhat simplified way the four most important deterioration mechanisms in reinforced concrete. As can be seen, the deterioration directly related to unsound aggregates is reduced in the example to alkali–aggregate reaction (AAR) and alkali–mineral reaction (AMR) and this is elaborated upon and discussed in more detail in section 6.4 onwards. The other deterioration mechanisms all relate in part to the physical performance of the aggregates (for

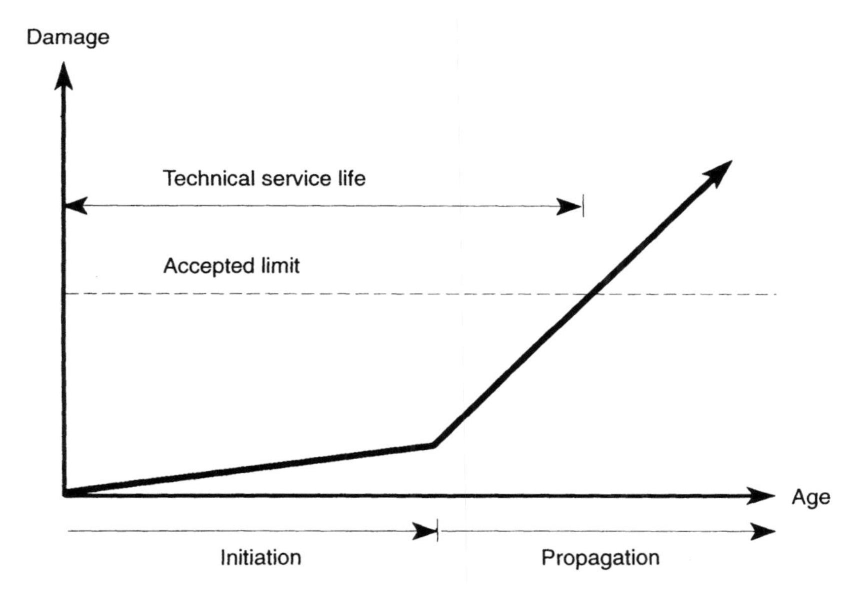

Figure 6.3 Progress of deterioration.

Table 6.3 Concrete deterioration related to aggregates, selected workmanship and environment factors and time[a]

Feature	*Possible effects*	*Time before effects may be observed*					
		hours	*days*	*weeks*	*months*	*years*	*decades*
1. Mainly workmanship							
Plastic shrinkage (including plastic settlement)	Cracking, localized loss of bond and/or cover to reinforcement	//////////					
Initial drying shrinkage	Crazing and/or cracking, reduced effective/actual cover to reinforcement		///	///	/////////////////////////		
Initial thermal contraction	Cracking in depth, localized loss of cover to reinforcement		///		////////////////////////		
Drying shrinkage	Cracking in depth, localized loss of cover to reinforcement			///	///	////////////////	
2. Aggregates							
Unsound aggregates generally	Surface disintegration, general loss of effective and actual cover to reinforcement, cracking				///	/// ///	////////////////
Reactive siliceous aggregates	Pop-outs, expansion and cracking, loss of effective and actual cover to reinforcement					/// ///	/////////////////

Reactive 'carbonate' aggregates	Expansion and cracking, loss of effective and actual cover to reinforcement	/// /// /// /// /// ///
Sulphate attack internal	Expansion and cracking, loss of effective and actual cover to reinforcement	/// /// /// /// ///
Excessive chlorides internal	Accelerated reinforcement corrosion cracking and spalling	/// /// /// /// //////////////////////
3. Environment		
Sulphate attack external	Expansion and cracking, loss of effective and actual cover to reinforcement	/// /// ///////////////
Excessive chlorides external	Accelerated reinforcement corrosion cracking and spalling	/// /// /// /// /// ////////////////////
Acid attack	Surface disintegration, general loss of effective and actual cover to reinforcement, cracking	/// /// /// /// /////////////////////////

[a] Source: Fookes, P.G. and Collis, L., *Concrete*, **10** (2), 14–19; 1976.

example, shape, porosity and strength); the quality of the concrete as indicated in section 6.1.2 is also related to the physical characteristics of the aggregates and the way they influence the mix design. Reinforcement corrosion may also be directly related to aggregates, e.g. if chloride were included in the mix as a contaminant from the aggregates.

The development of deterioration is such that aggregate-related problems are often slow to appear (but many of the problems may be seen much earlier under the microscope) and some of the quicker forms of cracking (e.g. plastic shrinkage or initial thermal contraction) are more directly related to workmanship.

Figure 6.3 is a simplification to highlight what is a common trait of many deterioration mechanisms. There is an initiation period during which, on a microscopic scale, events are happening within the concrete (e.g. ingress of chlorides or carbon dioxide; development of a gel in alkali–silica reaction) which take time to manifest themselves as visible forms of deterioration or cracking on the surface of the concrete. So many factors influence the rates of deterioration (including mix design, materials other than the aggregates, environment, exposure, use of the structure) that apart from the early features, particularly those related to workmanship like plastic shrinkage or initial thermal contraction, it is not realistic to put rigorous time scales on deterioration.

Having said this, Table 6.3, modified from Fookes and Collis [22], attempts to put time into the context of deterioration by a simple bar diagram showing some of the deterioration related to workmanship factors and also aggregates. This is inevitably a simplification and does not take fully into account aspects of design of the structure, mix design, environment, exposure and so on. The intermittent part of the time bar on the table indicates either an early extreme of the onset of the problem or the relatively minor nature of the problem.

6.2.2 Evaluation of deterioration

The evaluation of deterioration is again a growth area in interest and publications but, except where deterioration of aggregates is involved, is not the subject of this review. Nevertheless, in assessing the deterioration and performance of aggregates, it is essential to take a broad view at the start of the investigation so that the main possible causes and influences on deterioration not related to aggregates can be eliminated or at least brought into focus and the part played by the aggregates can be appropriately determined. Current British, American and other standards, textbooks (e.g. Kay, 1992 [23] and Mays (ed.) 1992 [24]) and current articles in concrete journals should be referred to for good practice in assessment of deterioration.

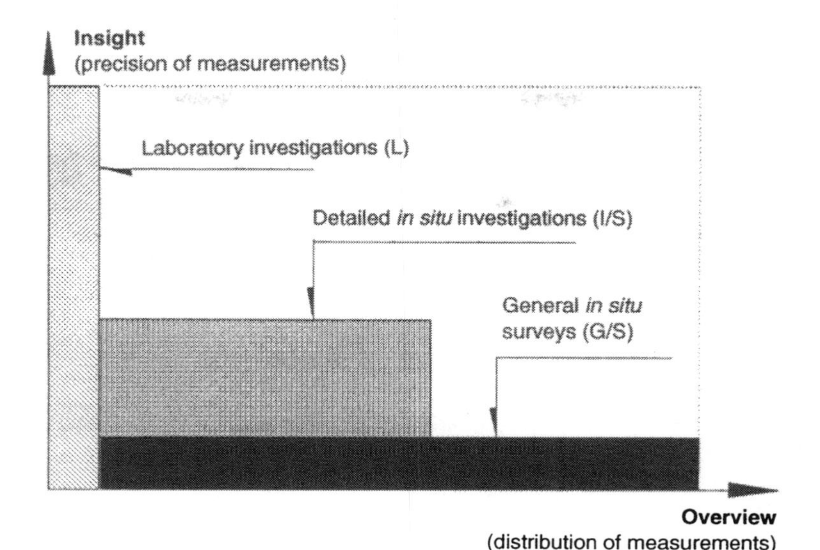

Overview in provided by, *inter alia*:
- History (G/S)
- Visual inspection (G/S
- Hammer tapping (G/S)
- Dynamic response tests (I/S)
- Test loading (I/S)
- Thermography (I/S)
- Half-cell potential mapping (I/S).

Insight is provided by, *inter alia*:
- Petrographic analysis (L)
- Carbonation testing (I/S)
- Chloride profile measurements (L)
- Radiography (I/S)
- Impulse echo (I/S).

Transition information is provided by, *inter alia*:
- Coring and core testing (L)
- Windows (I/S)
- Endoscope inspection of voids (I/S).

Figure 6.4 Information from different types of investigation. (Redrawn from Rostam, Philosophy of assessment and repair of concrete structures, and the feedback into new designs; 1991.)

Figure 6.4 gives a simple view of approaches to investigation of a structure, in part after Rostam [25], with examples of testing and techniques involved.

Table 6.4 is a checklist for the investigation of deteriorated concrete. All or some of actions 1, 2, 3 and 4 should invariably be carried out when starting an evaluation of a structure, even if a fairly specific form of aggregate problem is suspected as a prime cause (for example, ASR). The laboratory examination and tests (action 5) would depend to a large extent on the earlier findings (actions 1 to 4) of the evaluation.

Table 6.4 A checklist for investigation of deteriorated concrete

1. Concrete under inspection Sample(s) of concrete from a structure (sound and deteriorated) Laboratory specimen(s) stored on site Laboratory specimen(s) stored in laboratory Sampling procedure Sample storage and treatment A concrete structure	2. Initial data on concrete Concrete structure history (design, dimensions, loading) Concrete specifications Concrete mix design Tests on materials used Quality control of fresh concrete Quality of concrete in place Duration and type of curing conditions Age at time of attack
3. Influence from the environment Temperature Humidity Pressure Permeability of the surrounding media Sea water Other aggressive substances Type of contact Concentration of aggressive substances Frequency and duration of exposure Special environmental influences (e.g. stray currents)	4. Visual signs of deterioration Cracking Erosion Spalling Exfoliation Dusting Crumbling Softening Staining Pop-outs Liquid gel exudation Crystallization Corrosion of reinforcement Misalignment Others
5. Laboratory examination and tests Visual examination Chemical analysis Thermal analysis Infra-red spectrometry X-ray diffraction analysis Microscopy (optical or electronic) Mechanical tests	 Sonic tests Dimensional change Weight change Capillary absorption Permeability Porosity Others
Conclusions Design of concrete structure not appropriate Concrete specifications not appropriate Concrete specifications not fulfilled Control of concrete technology inadequate Unsatisfactory quality of components	Recommendations Safety precautions Demolish Repair Prevention

6.3 ROCK MATERIAL

6.3.1 Classification

Rock is strictly defined in geology as 'any natural solid portion of the Earth's crust which has recognizable appearance and composition'. Some rocks are not necessarily hard or strong and in discussion geologists may call peat or clay a rock as they would granite or limestone. This is a complex subject to which justice cannot be done here. However, for further reading, there are several excellent textbooks for engineers on the subject, of which Blyth and de Freitas [26] and Mclean and Gribble [27] are two of the most well known in the UK. The following short discussions and introduction to rock classification are given to enable the reader who is not familiar with the subject to follow a little better the discussion in section 6.4.

There are three major classes of bedrock:

1. *sedimentary rocks* formed by the deposition of material on the Earth's crust, e.g. shale, sandstone, limestone.
2. *igneous rocks* formed from molten rock magma solidifying either at the Earth's surface or within the crust, e.g. basalt, andesite, granite (*s.l.*).
3. *metamorphic rocks* produced deep in the Earth's crust by the transformation of existing rocks through the action of heat and pressure, e.g. marble, slate, gneiss.

In addition to aggregates produced from bedrock, a large proportion of aggregates come from most forms of alluvial and other superficial materials (mainly unconsolidated deposits, loosely called 'drift', e.g. clay, silt, sand, gravel, cobbles, boulders).

For more discussion on these aggregate sources, see Smith and Collis [2].

Data on the subdivision of rocks into the major classes and their description, distribution, occurrence, likely quality as aggregate and comments on them as aggregates are presented in the Appendix in Tables 6.A.1 to 6.A.6 and Fig. 6.A.1 and 6.A.2.

Aggregates from sand and gravel

Sand and gravel deposits are accumulations of the more durable rock fragments and mineral grains which have been released from their parent rock by physical weathering processes or abraded by the action of ice (e.g. glaciers) and then worn and sorted by water (e.g. rivers, waves) and the wind. Finer deposits (the matrix, e.g. mud, sand) often more or less fill the spaces between the larger particles (the clasts).

The principal sources of sand and gravel in the UK are mainly the geologically young unconsolidated sedimentary deposits which have accumulated in the relatively recent geological past, since the onset of the Pleistocene Ice Age, about two million years ago. In the UK, geological deposits, which form the bedrock ('country rock' or 'rock-head') upon which the softer or weaker drift materials rest, are grouped together by the British Geological Survey [28] as the 'solid' formations. Drift deposits include aeolian (i.e. wind blown), marine, beach and lacustrine deposits (i.e. the deposits of lakes), alluvial deposits (i.e. the deposits of rivers) and hill and mountain slope solifluction wastes (largely of periglacial freeze–thaw origin), as well as fluvioglacial deposits resulting from streams emanating from the Pleistocene ice sheets.

The properties of gravel, and to a lesser extent sand, largely depend on the rocks from which they are derived, although during long transport distances prior to their deposition, weathered or otherwise weaker fragments tend to be selectively worn away so that the remaining aggregate material is usually stronger than the crushed parent rock. On the other hand, where transport distances are short, for example in fluvioglacial environments or alluvial fans, deleterious constituents such as mud flakes, fragments of coal or chalk may remain and reduce the attractiveness of a prospective aggregate deposit.

Useful information, additional to that given below, on the nature and occurrence of these aggregate-bearing deposits, including guidance on their identification and description, can be found in the *Code of Practice for Site Investigations* BS 5930 [29], particularly section 7 and Appendix G.

Aggregates from the sedimentary rocks

As a result of the deposition of drift types of deposit on the land surface and their eventual burial, they form new sedimentary rocks. In addition, some sediments originate as chemical or biochemical precipitates and others as ash, dust and rock debris from volcanism (volcanoclastic deposits), and yet others, like peat and coal, principally occur as organic residues and are not derived from pre-existing rocks.

After long periods of geological time, the diagenetic geological processes associated with the burial of these sediments slowly cement (lithify or indurate) them to more or less strong rocks; for example deposits consisting of rounded clasts (e.g. boulders, cobbles or gravel), when cemented or lithified, are known as conglomerates (Fig. 6.5) – if composed of angular fragments they are called breccias.

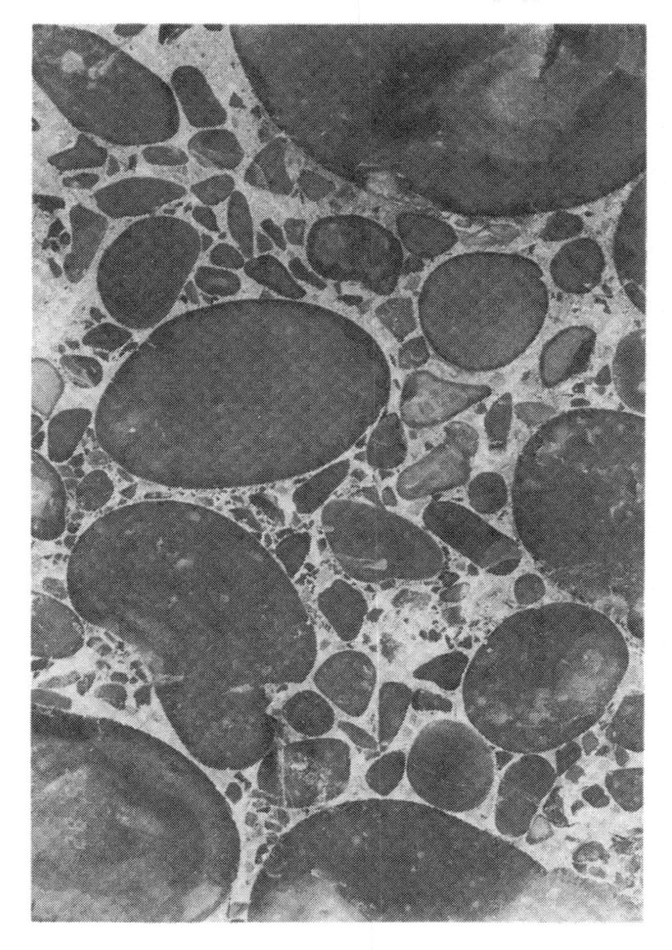

Figure 6.5 Conglomerate (Hertfordshire puddingstone), polished surface, width of photograph 80 mm. Large particles are rounded flints. (Reproduced with permission from Dr W.J. French.)

Some alluvial sands and gravels considerably older than the Pleistocene epoch (and classed as bedrock) may have had only relatively shallow burial (e.g. Oligocene sands and gravels in the Bovey basin, Devon) and are not yet strongly lithified and can be worked as though they were young alluvial sands and gravels.

Detrital (or clastic, i.e. composed of fragments) sedimentary rocks are generally classified on the basis of a convenient grain size, giving pebbly, sandy and muddy categories, and chemical and biochemical (organic) rocks on the basis of the predominant raw material.

Aggregates from igneous rocks

Igneous rocks (that is, rocks which have solidified from a fluid rock melt or magma) exhibit a very wide range of chemical composition which is reflected in the intricacy of their nomenclature and classification. Their suitability as aggregates depends on their mineral constituents, crystalline fabric (e.g. interlocking crystals), texture and degree of chemical alteration and weathering (if any). These in turn depend on their origin and subsequent geological history. Full discussions of their petrological characteristics can be found in appropriate geological textbooks.

The plutonic rocks formed at depth in large masses (e.g. granite, diorite) typically are of slowly cooled (i.e. solidified), coarsely crystalline texture; smaller bodies, which may have been intruded as dykes, sills or bosses into country rock, have a faster cooled, medium to fine crystalline texture (e.g. microgranite, dolerite), perhaps with a scattering of larger crystals (porphyries). Extrusive igneous rocks have solidified rapidly as lava flows on the land surface or under the sea and may be of glassy or microcrystalline texture (e.g. basalt, andesite). Lava flows may be of little use for aggregates if they include flow banded, strongly jointed, vesicular (slaggy) (Fig. 6.6) or brecciated (broken) material.

Fragmental (pyroclastic) materials, such as dust, ash, tuff and rock debris ejected from volcanoes, may not be useful unless they become either indurated into hard rock by heating (e.g. welded tuff) or compacted by burial or cementation (e.g. agglomerate); ancient volcanic and volcanoclastic rocks can yield high quality aggregate. Pumice, a highly vesicular, sometimes froth-like lava, resulting from the expansion of gases during solidification, is a naturally occurring lightweight aggregate favoured in the construction industry for making low-density concrete blocks for inner walls and cavity installation.

From a chemical point of view, the vast majority of igneous rocks are made up of combinations of only eight elements. Of these, oxygen is dominant, next is silicon and then aluminium, iron, calcium, sodium, potassium and magnesium. In terms of their oxides, silica (SiO_2) is by far the most abundant, ranging from 40% to 75% of the total. The proportion of silica in a rock can also be a guide to its hardness (i.e. approximating to its abrasiveness). It is the percentage of silica that forms the basis of a threefold classification into acid, intermediate and basic categories. Silica rich (acid) igneous rocks typically contain free silica in the form of clear, colourless quartz crystals and opaque, pale-coloured feldspars, which generally makes them paler and lighter in weight (e.g. granites). The intermediate (e.g. andesite) and basic (e.g. basalt) rock types have greater amounts of dark-coloured ferromagnesian minerals of higher density and only very rarely any free quartz.

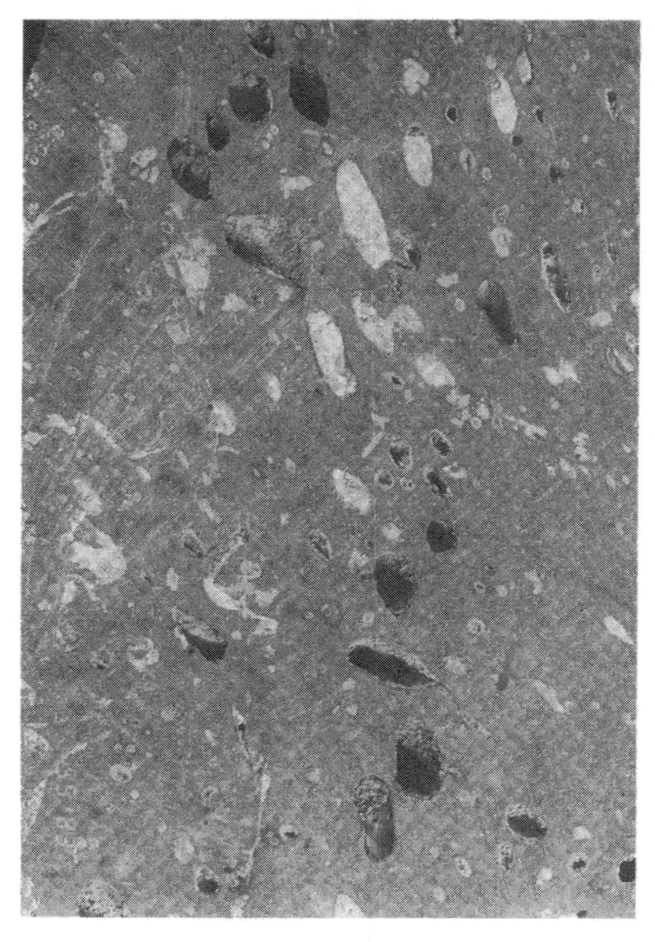

Figure 6.6 Basaltic andesite with vesicules (gas bubble holes) infilled with smectite (swelling clay). Width of photograph 40 mm. (Reproduced with permission from Dr W.J. French.)

Aggregates from metamorphic rocks

Metamorphic rocks result from the alteration of existing igneous, metamorphic and sedimentary rocks by heat, pressure and chemical activity, and comprise a very diverse and complex range of rock types. Their usefulness for aggregates is equally variable. The massive (i.e. without joints) metamorphic rocks probably provide the best prospective aggregate sources, although some may be too strong (e.g. some quartzites) to be worked economically for undemanding applications.

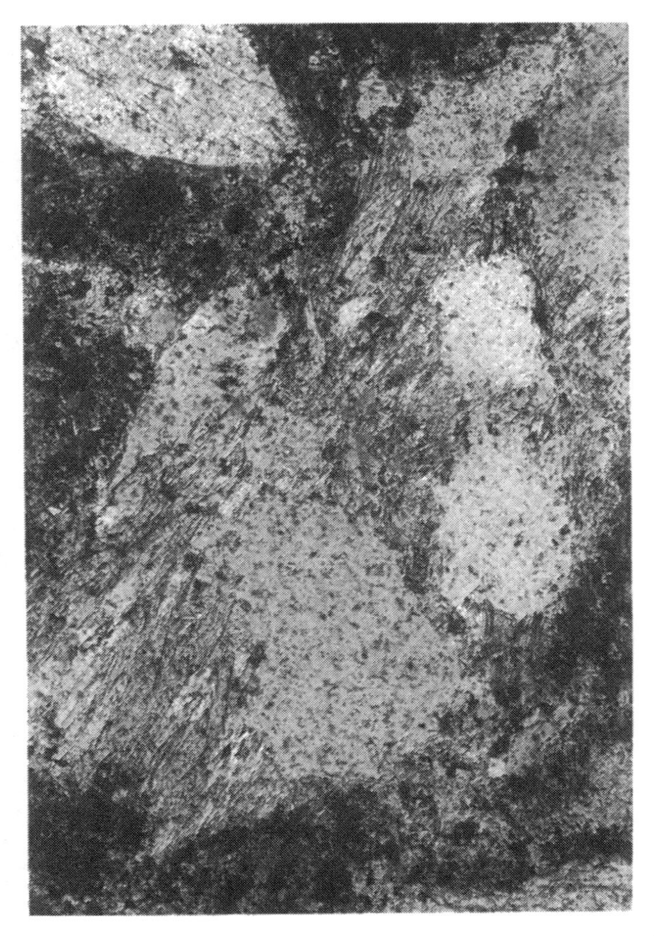

Figure 6.7 Sand grain of mica schist with orientation of mica grains. Alignment of mica bottom left to top right. Width of photomicrograph 1 mm. (Reproduced with permission from Dr W.J. French.)

Foliated rocks and rocks with cleavage may be less attractive as they may break to give poor shapes (e.g. flaky, elongate) or split apart in production or service. Aligned elongate rock fragments may induce anisotropic weakness in concrete.

Thermal or contact metamorphism takes place when rocks are heated by the intrusion of a mass of molten igneous rock into clay rocks, such as shales and mudstones, which may be recrystallized into hard, dense, flint-like rock, termed hornfels. Quartz-rich rocks, the greywackes and sandstones, fuse or recrystallize into quartzites and psammites which can provide high quality aggregates.

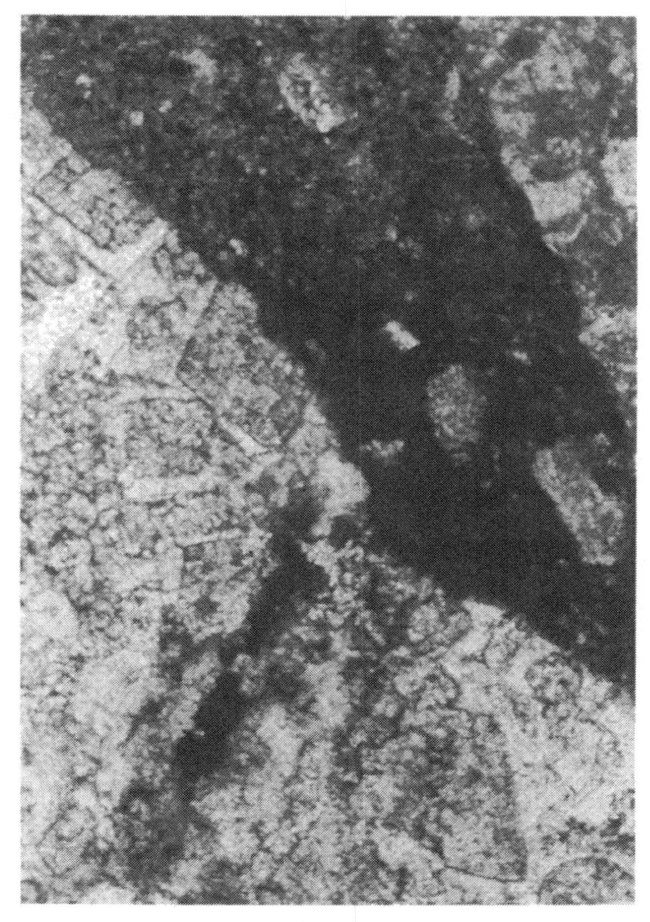

Figure 6.8 Coarse aggregate of phyllite with orientated mica and other mineral grains. Alignment of grains bottom left to top right. Width of photomicrograph 1 mm. (Reproduced with permission from Dr W.J. French.)

The effect of progressive regional dynamic metamorphism (that is, the widescale alteration induced by extreme pressure and heat through the deep burial of rocks in the crust) is firstly to impart foliations; the foliations may not always be visible to the naked eye, although individual mineral grains may exhibit strain lines when seen under the microscope. When this happens to a quartz-rich rock, it may become potentially reactive to alkalis. Where local pressure-induced movement occurs, the rocks become sheared. Plated or foliated structures in metamorphic rocks indicate that high shearing stresses have been a principal agency in their formation.

More intense metamorphism gives rise to widespread recrystallization of minerals and in extreme conditions many rock constituents melt and become intruded as an igneous body. Under these conditions, clay-rich sediments first form slates, which are characterized by closely spaced cleavage planes perpendicular to the direction of greatest pressure, then phyllites (foliated rocks containing much white mica) and finally schists and pelitic gneisses, in which the constituent minerals tend to be irregularly aligned in partings with an uneven, wavy schistose texture.

Gneisses are formed from igneous, metamorphic and sedimentary rocks in more extreme conditions. They are typically coarse grained and banded in structure, reflecting differences in original composition accentuated by segregation of the constituents during recrystallization of flow under great pressure deep in the Earth's crust. Striped or injection gneisses may result from the injection of thin sheets of molten material (often rich in quartz and feldspar) along parallel planes in a variety of country rocks; migmatites are generally formed by partial melting and recrystallization of the host rock. Such rocks are often massive (i.e. without discontinuities) and can make good quality aggregate.

6.3.2 Rock masses

All rock masses are fractured by discontinuities in one or more of a variety of ways and often it is these fractures which control the permeation of air, fluids or gases which may ultimately alter the rock by weathering or other processes. In the winning of the rock in quarries, discontinuities affect blast waves and help ripping of otherwise strong indurated rock. When the rock is processed as aggregate, fractures help control, often on a microscopic or submicroscopic scale, the physical characteristics of that aggregate. Again, this is a large subject and the textbooks should be referred to for more details.

Table 6.5 gives the weathering grades generally adopted by British Standards. Rock weathering can be very important in affecting the characteristics of rock and aggregates made from weathered (or altered) rock. Although often thought of as an overseas problem, weathered rock from former subtropical climates and periglacial activity (during the ice ages) does exist in places in the UK. Some of the rock in south-west England, and parts of the English Midlands and Scotland in particular, can be described as 'highly' weathered (BS 5930 weathering grade IV). The effect of the degree of weathering on geomaterials, especially aggregates, broadly indicated by the weathering grade, is also given in Table 6.5, after Fookes [30].

Table 6.5 Weathering and alteration grades of rock, and estimated characteristics of aggregate

Weathering definition	Grade		Estimated aggregate characteristics
		Humus/topsoil	
All rock material converted to soil; mass structure and material fabric destroyed. Significant change in volume	VI	Residual soil	May be suitable for random fill.
All rock material decomposed and/or disintegrated to soil. Original mass structure still largely intact.	V	Completely weathered	Not suitable for aggregate or road pavement but may be suitable for select fill.
More than 50% of rock material decomposed and/or disintegrated to soil. Fresh/discoloured rock present as discontinuous framework or corestones.	IV	Highly weathered	Not generally suitable for aggregate but may be suitable for lower parts of road pavement and hardcare.
Less than 50% of rock material decomposed and/or disintegrated to soil. Fresh/discoloured rock present as continuous framework or corestones.	III	Moderately weathered	Aggregate properties will be significantly influenced by weathering. Soundness characteristics markedly affected. Alteration of mineral constituents common and much microcracking.
Discoloration indicates weathering of rock material and discontinuity surfaces. All rock material may be discoloured by weathering and be weaker than in its fresh condition.	II	Slightly weathered	Aggregate properties may be significantly influenced by weathering. Strength and abrasion characteristics show some weakening. Some alteration of mineral constituents sound.
Discoloration on major disconuity surfaces.	IB	Faintly weathered	Aggregate properties not influenced by weathering. Mineral constituents sound.
No visible sign of rock material weathering.	IA	Fresh	Aggregate properties not influenced by weathering. Mineral constituents of rock are fresh and sound.

Idealized weathering profiles– without corestones (left) and with corestones (right)

Example of a complex profile with corestones

Key:
- Rock decomposed to soil
- Weathered/disintegrated rock
- Rock discoloured by weathering
- Fresh rock

6.3.3 Investigation of rock as potential aggregate

Natural aggregates are rarely if ever composed of one uniform lithology. This obviously has important consequences for the potential occurrence of deleterious materials and for the investigation of these materials.

A number of minor constituents can prove to be deleterious (e.g. through alkali–aggregate reaction or through the presence of sulphides) where the main lithology is regarded as highly satisfactory and inert. An example of this is that most gabbros would be regarded as ideal lithologies for aggregate and innocuous from the point of view of alkali–aggregate reaction. One gabbro, however, because of a local minor ingredient, gave rise to the first serious deterioration recorded through alkali–aggregate reaction in the UK, on a concrete dam in Jersey, Channel Islands [31].

French [1] suggests that a more than usually 'homogeneous' granite quarry, for example, could well be expected to include any of the following, some of which might turn an otherwise sound rock into an unsound aggregate.

- Compositional variations of mineral proportions and mineral species;
- Textural variations of grain size, mineral intergrowth, mineral alignment and so on;
- Hydrothermally altered zones, patches, joints, and specific lithologies;
- Weathered zones, including mineral changes, sheet silicas, oxides and sometimes sulphates and carbonates and rock material weakening;
- Mineralized zones and veins sometimes including special silicates, sulphides, fluorides, carbonates, sulphates, boron-bearing phases, and other minerals of special composition;
- Dykes and other sheets of contrasting lithology and texture such as microgranites, felsites, porphyrites, basalts, microdiorites, etc.;
- Enclaves of metamorphic rock, i.e. floaters (inclusions);
- Sheared rock resulting from faulting or small movements along joints;
- Cataclasite resulting from localized or general movement along shear planes producing granulation and even fusion of the parent rock type;
- Tuffisite, usually a compact, granular fine-grained rock arising from the passage of gases through the granite.

As a simple example of this, Fig. 6.9, based on field notebook sketches of a real quarry in a Permian–Carboniferous sill in the Midland Valley of Scotland (Pettifer, G. personal communication), shows an elevation of a quarry face. It has been drawn twice: (a) as likely to be seen by the quarryworker, and (b) as likely to be seen by the geologist.

Variations expected for each rock type are mostly well known to geologists and can be given for each intended lithology or aggregate

Table 6.6 A systematic procedure for the examination and testing of potential aggregate sources[a]

A. Reconnaissance of the available materials and establishment of their performance in likely applications

1. Establish the geological basis for the investigation by means of a desk study to collate all relevant data and undertake a walk-over field survey, including any special geological investigations to establish existence of potential sources.
2. Undertake field sampling of likely sources; investigate the characteristics of any existing nearby aggregate sources, including the range of end-uses of the products, whether or not they are similar to those for which the present appraisal is being conducted.
3. Carry out laboratory examinations and tests on samples obtained during steps 1 and 2 above; assess the performance of existing locally available aggregates in use; for example, the performance of aggregate **within** concrete as well as of the concrete itself.

B. Examination of potential aggregate sources – site investigation

1. Carry out full-scale trenching, pitting and/or drilling and sampling of the potentially useful occurrences identified from investigations so far, including material from existing local aggregate workings, whether or not it is intended for use.
2. Carry out laboratory investigations and tests on the samples to determine their physical characteristics and potential suitability; relate this information to the known geology of the rock mass(es) under investigation and obtain an indication of the size and distribution of the potential mineral reserve.
3. In the light of all the new information assess the extent to which selective extraction and special processing techniques may be required to ensure a bulk supply of aggregate of sufficient quality. Review the need for particular processing plant designs.

C. Aggregate production trials

1. Plan and supervise materials extraction and aggregate production trials including representative sampling of end products, selective sampling of both accepted and rejected material.
2. Carry out further laboratory tests on samples collected from the extraction and production trials, review and finalize extraction methods, production procedures and target qualities for the processed aggregates for approval purposes.

D. Production control programme

1. Establish, approve and implement sampling and testing regimes for production control (producer) and specification compliance (engineer) purposes.
2. Devise, approve and maintain data records, reporting proformas and control charts to monitor aggregate quality and variability.

[a] Source: Smith, M.R. and Collis, L. (eds), *Sand, Gravel and Crushed Rock Aggregates for Construction Purposes*, 2nd edn; Geological Society, 1993.

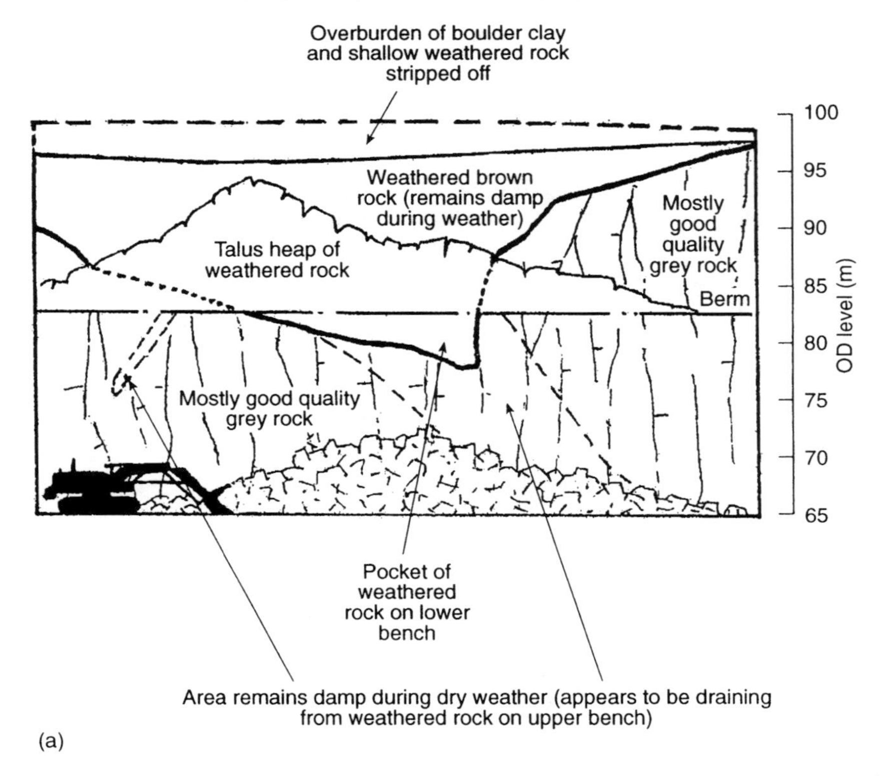

Figure 6.9 Field notebook sketches of a quarry face (a) general appearance of face (b) detailed geological map of face. (Source: G. Pettifer.) The rock is mainly quartz–dolerite. Magmatic differentiation has resulted in the local development of less basic quartz-diorite, which in turn has zones of intense late-stage hydrothermal alteration.

type. The approach to aggregate selection must therefore begin by a definition of:

- the physical and mechanical qualities required for the concrete;
- the material properties that are considered to be unacceptable.

Both these factors are dependent on the range of rock lithologies (i.e. the rock types) likely to be present in the potential quarry and it is essential that the selection begins by a study of the quarry. This review of the quarry may need to continue throughout the development of the construction project.

It is clear, therefore, that the proper study of the potential aggregate starts with the greenfield site or, if it is an existing aggregate, in the actual quarry or pit.

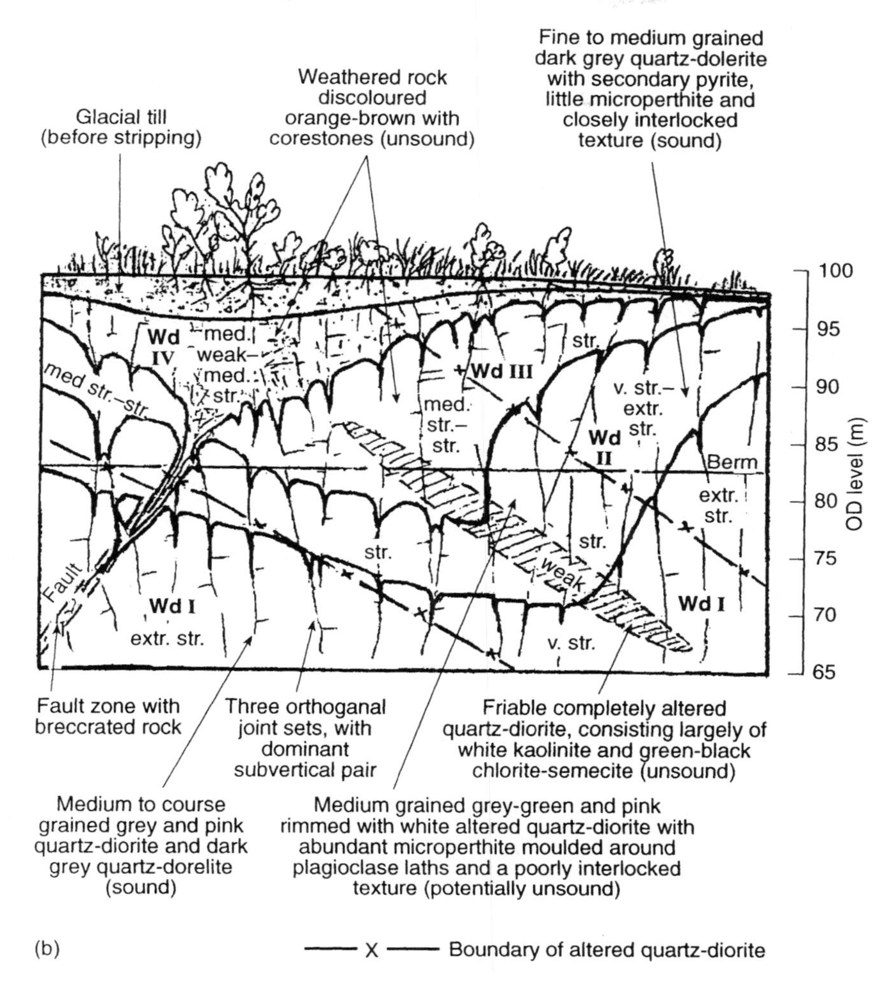

Figure 6.9(b) (*continued*)

Table 6.6 gives a systematic procedure recommended by the Geological Society working party report on aggregates [2]. The original source should be referred to for detail.

If the aggregates exist and are required to be fully evaluated, Table 6.7 (after French [1]) gives an outline of the steps which could be followed in assessing the aggregate qualities.

A selection of laboratory tests for general and specific evaluation of aggregates is given in section 6.4.1.

Table 6.7 Assessment of aggregate qualities[a]

A. Desk study

1. Consider proposed mix design, potential ingredients and the expected use of the concrete.
2. Estimate potential alkali levels from all sources.
3. Assess the environment of the concrete and the durability requirements.
4. Assess possible aggregate sources and any lithological characteristics likely to be problematic.

B. Field study

1. Make a walk-over study of the source, a field examination of stockpiles, and assess the detailed work needed.
2. Evaluate the types, distribution and abundance of rock types in the source.
3. Study crusher output at the hand specimen size (> 50 mm) and record rock types and proportions present.
4. Take rock samples to represent the range of lithologies present – in some cases large enough to make a trial aggregate of a specific lithology.
5. Examine and sample existing stockpiles and/or obtain conveyor belt samples.

C. Laboratory work

1. Examine samples and aggregate in thin section. Consider crushing rock or aggregate if this will make the sections more representative. Seek evidence of features indicating a potential for damage of all sorts.
2. Consider making trial aggregates and use them and field aggregates for gel pat and/or accelerated mortar bar tests – preferably the latter.
3. If ambiguity remains and time permits, carry out concrete prism tests using aggregate taken from the source.

[a] Source: French, W.J., in *Improving Civil Engineering Structures – Old and New;* Geotechnical Publishing, 1995.

6.4 PREDICTION AND PERFORMANCE

6.4.1 Unsatisfactory physical and chemical characteristics of some aggregates

Many characteristics of aggregates can be related to their influence on the performance of the concrete, i.e. the workability and curability when it is being manufactured, and dimensional stability, i.e. its soundness, during its service life.

The UK is fortunate in generally having ample supplies of sound natural aggregates for concrete, from major rock sources such as limestone, sandstone, granite, basalt, dolerite, flint gravel and silica sand. This is not always the case in other parts of the world, for example in arid coastal regions, where many local aggregates are contaminated with soluble salts, in particular sulphates and chlorides, both of which can react adversely in concrete. The presence of sulphates can lead to chemical attack on the aluminates of Portland cements, while chlorides can cause corrosion of reinforcing steel. Both reactions are expansive and result in cracking and disruption of the concrete.

In the UK, some relatively minor defects, such as surface disfigurement caused by staining or popping, can result from chemical reactions between cement and aggregate constituents, but in general no structural damage is caused. However, and more importantly, chemical reactions may also occur between certain mineral constituents of aggregates and the hydroxides of the alkali metals, sodium and potassium, present in the cement, resulting in cracking of the concrete.

Some UK rocks exhibit marked shrinkage on drying, and when used as concreting aggregate can cause cracking [20], but this is a physical property of the rock and chemical reactions are not involved. Undesirable impurities, whose effects are also physical, include excessive numbers of shells in marine-dredged aggregates, mica in concreting sands, and weak materials such as clay lumps or coal, and limits have been given for their permissible contents [32, 33].

Even though this chemical and physical unsoundness is relatively rare in the UK, a careful assessment of the aggregate should be made for each specific concrete structure in its local environment (sections 6.2.1 and 6.2.2). In a country like the UK, with a long history of winning of concrete aggregates, the track record of the particular aggregate is a good starting point for its assessment. Guidance can also be obtained, for example, from some of the older versions of BS 882 [18], in the Concrete Society Technical Report 30 (the report of the Hawkins Committee) [34] and in the DoT Specification for Highway Works [35] on possibly unsatisfactory materials in aggregate for use in concrete.

Section 6.3.3 discussed the investigation of rock sources as potential supplies of aggregates. The main features of importance in the prediction of likely performance of an aggregate are:

- rock types present and their abundance, including contaminants;
- properties of the rocks, mineralogy and texture, density, water absorption and strength;
- size range and shape of particles;
- nature of the interface between aggregate and binder;
- internal cracking in the aggregate and any associated cracks in the binder.

Rock characteristics which might prove unsatisfactory to a greater or lesser extent include those described in the following sections.

Physical features

As discussed in section 6.1.2, it is generally well known that the grading and physical characteristics of an aggregate are important factors influencing the workability, strength and economy of concrete.

Flaky particles make for a harsher mixture of lower workability at a given water content than non-flaky ones. This leads to poor compaction and a high void content resulting in low strength and durability. In particular, the depth of penetration of carbonation is largely governed by the degree of compaction so that poor aggregate shape can therefore influence reinforcement corrosion. Rocks that give rise to excessively elongate or flaky fragments often include gneiss, schist, amphibolite, phyllite and slate and those containing much coarse mica.

The angularity of the aggregate has an even greater effect on strength and workability than the flakiness index *per se* [36]. It is measured by the method given in BS 812 [37] in which the aggregate is packed in a standard volume. Under the microscope two aspects of angularity are apparent: the degree of rounding at edges and corners and the angles defined by the surfaces making those angles and corners. It is the latter which has the greater influence on aggregate packing and the harshness of the aggregate, while the former is likely to affect the location of microcrack development.

Microcracks are commonly developed on the surfaces of coarse aggregate particles (Fig. 6.10). These may also occur along the surfaces of fine aggregate grains if thermal stresses are high, as, for example, might occur when concreting in cold weather (Fig. 6.11). The development of some cracks at this location is not unusual even in concrete that is well made and of considerable strength. It appears to represent early hydration shrinkage and is often the focus of the formation of chains of portlandite crystals. Cracks of this type are commonly encountered even when the water/cement ratio of the mixture is as low as 0.4.

It is very common for cracks and microcracks to develop along the interface between the aggregate and the cement paste and there is often a difference in the coefficient of thermal expansion between the aggregate and the binder (Fig. 6.12). Stresses are therefore commonly set up along these zones. In the case of siliceous aggregate, layers of calcium hydroxide (portlandite) frequently develop along the aggregate surfaces, with their form and abundance depending upon the composition of the binder and the conditions of curing.

Aggregates may influence the role of water in the binder or allow penetration of moisture and the atmosphere deeper into the concrete than is desirable (e.g. porous limestones, sandstones and mudstones). It is therefore important to consider aggregates which may:

- have a high microporosity, i.e. interconnected pores less than 3 μm in diameter (the smaller the worse);
- affect the ultimate strength of the concrete, e.g. contain swelling and/or shrinking minerals, especially swelling sericites, smectites (e.g. some mudstones), weak particles, or flaky minerals such as mica;
- contain or be liable to produce dust (e.g. many limestones);
- lead to surface damage or more penetrative cracking due to freeze–thaw processes (e.g. porous sedimentary rocks).

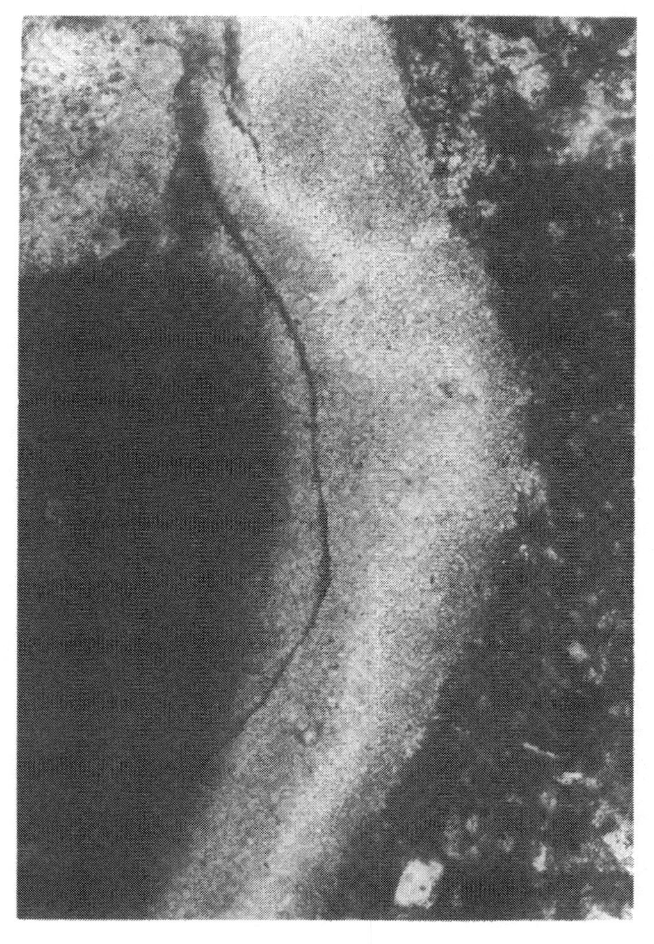

Figure 6.10 Microcrack in chert coarse aggregate parallel and near to surface. Width of photomicrograph 0.5 mm. (Reproduced with permission from Dr W.J. French.)

Figure 6.11 Microcrack at the paste – sand grain interface. Width of photomicrograph 1 mm. (Reproduced with permission from Dr W.J. French.)

Mineral features

The range and complexity of problems in concrete related to the mineralogy of the aggregate is discussed more fully in section 6.4.3 and can be referred to generally as aggregate–mineral reaction (AMR); alkali–aggregate reaction (AAR); or alkali–silica reaction (ASR), which is a special case of AAR that can lead to particularly extensive cracking. In summary, potential problem aggregates are:

ASR:

- those containing siliceous particles less than about 5 μm in grain size or including some glass (e.g. fine igneous lavas and fine metamorphic rocks) which can lead to alkali reaction;
- those containing alkali-reactive minerals such as chalcedony, tridymite, cristobalite, opal;

AAR:

- possibly finely divided sheet silicates such as chlorite and vermiculite, and rocks such as argillites, meta-argillites, greywackes and meta-greywackes;

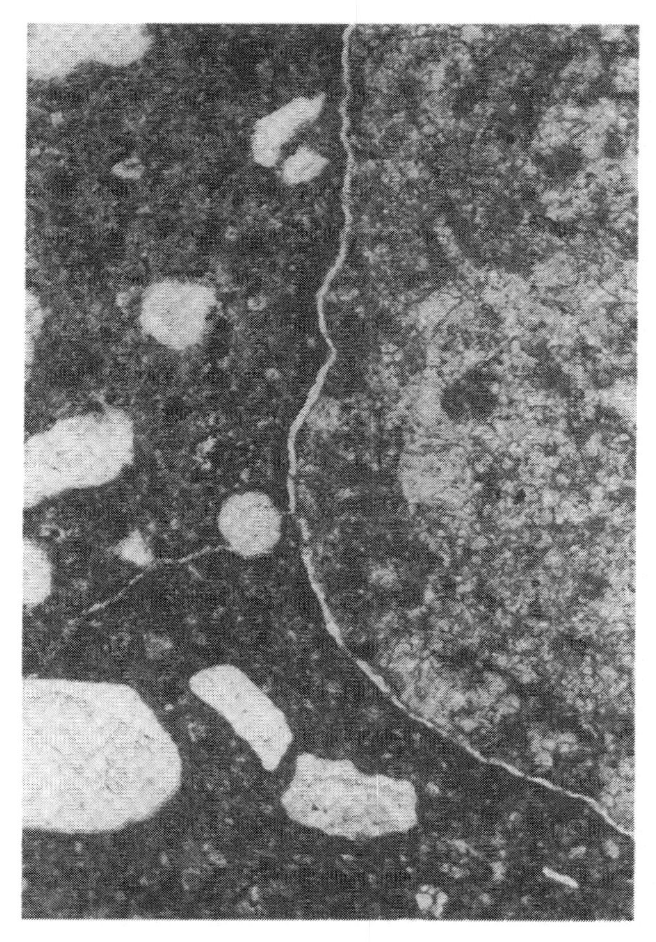

Figure 6.12 Microcrack at the paste–coarse limestone aggregate interface, limestone on the right. Width of photomicrograph 1 mm. (Reproduced with permission from Dr W.J. French.)

- carbonate rocks containing dolomite and fine-grained siliceous phases (typically some common non-clastic sedimentary rocks) which can lead to alkali–carbonate reaction (also called ACR);

AMR:
- those containing sulphides and sulphates, especially pyrite, pyrrhotite, marcasite, gypsum and anhydrite;
- those containing bituminous residues;
- contaminant minerals in aggregate which can adversely affect the hydration of the cement paste (e.g. organics, chlorides, phosphates);
- those containing certain zeolites (e.g. some igneous lavas), which can show volume changes or release alkalis into the cement paste (for reaction with other ASR or AMR minerals).

Weathering features

These can lead to physical or chemical unsoundness, e.g.

- weathered rocks (many types) containing oxides which can be dispersive and staining (AMR);
- weathered rocks (e.g. some igneous rock in Scotland and south-west England; some mudstones) containing secondary swelling and/or shrinking clay minerals (e.g. swelling sericites, smectites), Table 6.5;
- weathered rocks (many types) with increased numbers of micro-fractures;
- weathered rocks which have lost some or all of the cement that binds their grains together (e.g. a carbonate-bound sandstone).

6.4.2 Testing for physical and chemical unsoundness

Table 6.8 summarizes commonly used test methods for the assessment of aggregate durability, i.e. for physical and chemical unsoundness. Table 6.9 lists the main petrological features of the aggregates which affect physical test results. Table 6.10 gives tests for the specific evaluation of alkali–aggregate reactions.

Table 6.9 is a simple matrix that allows a quick and rough assessment to be made of the suitability of the test or engineering property to help quantify the important petrological properties which relate to the aggregate soundness in question.

A great deal of work has been carried out on the development of test procedures for the determination of the potential for alkali–aggregate expansive reactions. Work carried out over the last few years in Canada has been of particular importance [37, 56]. As a consequence of the difficulties encountered with some of these tests, interest has centred upon the development of tests akin to the NBRI mortar bar and concrete

Table 6.8 Commonly used test methods for the assessment of concrete and aggregate durability

Physical tests	*Mechanical tests*	*Simulation tests*	*Chemical and petrographic evaluation*
Aggregate grading (BS 882, BS 812, ASTM C33 and 136) [18, 37, 33, 38][a]	Point load strength (ISRM) [42]	Modified aggregate impact value (Hosking and Tubey) [45][e]	Petrographic examination (ASTM C295) [51]
Aggregate shape[b], angularity[c], sphericity, roundness, surface texture[d] (BS 812) [37]	Strength: – Aggregate impact value (BS 812) [37][e,f] – Aggregate crushing value (BS 812) [37][f] – Ten per cent fines value (BS 812) [37] – Point load test (Franklin) [43][g] – Schmidt rebound hammer (Duncan) [44][h]	Slake durability (Franklin and Chandra) [46]	Clay mineral analysis (XRD, DTA, methylene blue absorption) (Smith and Collis) [2]
Relative density, bulk density, unit weight (BS 812, ASTM C33 and 127) [37, 33, 39]		Freeze–thaw durability test (AASHTO T 103–78, DIN 4226) [47, 48]	Chloride content (BS 812) [37][m]
Water absorption (BS 812, ASTM C127 and 128) [37, 39, 40]		Sulphate soundness (ASTM C88) [49][i]	Sulphate content (BS 812) [37][n]
Aggregate shrinkage (BRE Digest 35) [20]		Mechanical and physical durability: – Aggregate abrasion value (BS 812) [37][j] – Aggregate attrition value – wet and dry (BS 812) [37][k] – Los Angeles abrasion value (ASTM C131) [50] – Polished stone value (BS 812) [37][l]	Organic content (BS 1377) [52]
Sand equivalent value (ASTM 2419) [41]			

Table 6.8 *(continued)*

[a] In engineering usage the terms describing sediment grading differ from those in geological usage, e.g. a sediment containing a range of sizes from fine to coarse would be classified as well graded in engineering terms but the same term would mean poorly sorted to a sedimentologist.
[b] In both natural and crushed rock aggregates the constituent particles within a particular size fraction may display a range of shapes. The distribution of shapes reflects intrinsic petrological–petrographic characteristics, together with environmental factors in formation (natural gravels) and production factors (crushed-rock aggregate). The most generally desirable aggregates are those with a high proportion of roughly equidimensional (cuboidal) particles.
[c] This is a measure of the lack of rounding of aggregate particles. This characteristic adversely affects the workability of concrete but can improve the stability of interlocking fragments.
[d] This expression of the microtopography of aggregate particles reflects petrographic and environmental factors of the original material. This property affects frictional properties and intergranular slip in unbound aggregate and adhesion.
[e] Two tests are performed and the results should be within a numerical value of 1. In this test a lower numerical value indicates a more resistant rock. The reproducibility of results in the test is high so that two values per sample are sufficient. The apparatus used in this test is relatively portable and cheap to operate, allowing both laboratory and field testing of aggregate.

Hosking and Tubey [45] developed a modified version of the BS 812 impact test and recent work has found the test particularly useful in discriminating variously weathered aggregate materials, particularly in the difficult boundaries of grade I–II and II–III. The nature of the test is such that it simulates dynamic loading of aggregate in a saturated state.
[f] These tests do not measure strength in the sense of compressive strength of intact rocks, rather they are indices of the resistance to pulverization over a prescribed sequence of loading. The variation in aggregate impact and crushing values within specific rock type categories, or between broader rock groups, can be attributed to the influence of aggregate particle shape [53] and geological features such as bulk composition, e.g. silicate and carbonate rocks, etc.; grain size; texture and structure; alteration.
[g] The Franklin point load apparatus is a simple, portable device for obtaining an indirect measure of the compressive strength of rock cores in the field. It can be used, however, with small irregular lumps as might be obtained from the ledge rock or quarried rock.
[h] An experienced geologist can obtain a qualitative impression of the toughness, elasticity and state of freshness of a rock from the impact and sound of his hammer striking the rock. The Schmidt hammer is a simple quantitative extension of this test. Although originally designed as a non-destructive test for concrete, it has been applied unmodified to natural rock materials. The results obtained, however, can be strongly influenced by test method factors such as surface preparation for the plunger and the unsuspected presence of open, face-parallel cracks close behind the rock surface.
[i] This subjects the sample of aggregate to the disruptive effects of the repeated crystallization and rehydration of magnesium (or sodium) sulphate within the pores of the aggregate. The extent of the disruption is dependent upon the physical soundness of the aggregate. The degree of degradation arising from the disruptive effects is measured by the extent to which material finer than 10 mm in particle size is produced. The test is carried out in duplicate and the mean result to the nearest whole number is taken as the magnesium sulphate soundness value.

Crystallization tests in general, and the magnesium sulphate test in particular, are severe tests. Such tests are known to give different results depending on the nature of the crystallization medium and the number of cycles involved. Also, the shape, size, porosity and permeability of the aggregate particles have significant effects. Commonly in practice these tests may measure the number of friable particles among otherwise sound aggregate rather than the general performance of a uniform aggregate. Thus these tests are only applicable in special circumstances where an exceptional minimum of particle deterioration is being sought. The normal applications in the UK involve aggregates used in the surface of airfeild runways, taxiways and hardstandings, as well as motorways and trunk roads.

[j] The AAV reflects hardness and brittleness of the constituent minerals, the influence of mineral cleavage and the strength of intergranular bond. Igneous and non-foliate metamorphic rocks yield the highest values within comparable composition ranges. Aggregate abrasion value is related in a simple and rational fashion to AIV and PSV for arenaceous rocks.

[k] This test, sometimes referred to as the 'Deval' test, is little used in the UK except for railway ballast and is no longer in BS 812. Reproducibility is poorer than other aggregate tests including the Los Angeles test, which is more widely favoured.

[l] The polished stone value was designed as a predictive measure of the susceptibility of a stone to polishing, when used in the wearing surface of a road.

[m] A test method for measuring the amount of water-soluble chloride present in aggregate, based on that of Volhard where an excess of standard silver nitrate solution is added to the chloride solution acidified with nitric acid, and the excess silver nitrate is back-titrated with potassium thiocyanate using ferric ammonium sulphate as an indicator.

Other methods are the Mohr direct titration method in neutral solution, using potassium chromate as an indicator; measurement of chloride concentration by specification electrode or, for field work and production control work of saline aggregates, the use of test strips [54].

The method given in BS 812, and that of Figg and Lees is designed particularly for testing those aggregates, such as marine-dredged flint, where the chloride is a surface contaminant only. With porous aggregates, or those of sedimentary or evaporite origin, where the chloride can be disseminated throughout the particles, complete water extraction may not be achieved unless the aggregates are finely ground before test.

[n] The sulphate content usually refers to the total acid-soluble sulphate content expressed as a percentage of SO_3 and methods for carrying this out are included in several British Standards for testing aggregates and related materials (e.g. BS 812 [37]; BS 1047 [55], BS 1377: pt 3 [52]). All the methods involve the extraction of the sulphates with hydrochloric acid and the gravimetric determination of the sulphate ions by precipitation with barium chloride.

The determination of the total sulphate content can give a false impression of the potential damage to concrete since if all the sulphate is present as the sparingly soluble calcium sulphate, risk of sulphate attack is much reduced because of its low solubility. For this reason several of the standards include a method that involves the extraction of the sulphate ions with a limited amount of water (usually two parts by mass of water to one part of aggregate) because this limits the importance attached to calcium sulphate. If calcium sulphate is the only sulphate present then water extraction can only give a value for sulphate content up to 1.2 g/l SO_3 as this is the solubility of calcium sulphate. Values in excess of 1.2 g/l must indicate the presence of soluble and potentially more harmful sulphates.

Table 6.9 Main petrological features affecting engineering test results[a]

	Aggregate characteristics												
Engineering test or property	*Specific gravity*	*Mineralogy*	*Compactness*	*Porosity*	*Hardness*	*Texture*	*Shape*	*Moisture content*	*Anistropy*	*SM content*	*Microfractures*	*Cleavage*	*Grain size*
Water absorption			x	x						x	x		
Specific gravity	x	x	x	x						x	x		
Compressive strength		x		x	x		x	x	x	x	x		
Tensile strength			x			x		x	x	x	x	x	
Shore hardness value		x	x	x	x			x			x		
Slake durability	x			x	x	x		x		x			
Washington degradation test	x	x	x	x			x		x				
Wetting and drying				x		x				x	x		
Freeze–thaw			x	x		x		x			x		
Sulphate soundness			x	x			x		x	x	x		
Rapid abrasion test	x		x		x	x	x	x			x		x
Los Angeles abrasion	x		x		x		x				x		x
AAV			x		x	x	x				x		x
ACV		x	x	x	x					x	x		
10% fines			x		x	x	x				x	x	x
PSV		x			x	x	x			x		x	x
Aggregate shrinkage		x		x				x		x			
Sonic velocity			x	x				x			x		
Modified AIV	x		x	x	x	x	x	x		x	x		
Ultrasonic cavitation			x		x	x				x	x	x	

[a] Source: Fookes, P.G. *Quarterly Journal of Engineering Geology*, **24**(1), 3–25; 1991.

prism tests. Concrete prism tests are at first sight most appealing because they can test the aggregate in the size ranges that are to be used. However, there are some difficulties with these tests in that the reaction may be dependent upon the quality of the concrete produced. Table 6.10 summarizes the tests in common use in the UK and elsewhere.

6.4.3 Aggregate changes within concrete during its lifetime

One of the most obvious uses of the petrographic investigation of concrete is the observation of changes that occur within the hardened concrete. The large majority of rocks used as aggregate remain

Table 6.10 Common tests for evaluation of alkali–aggregate expansion potential

NBRI mortar bar test (BS 812 [37] and Oberholster and Davies [57]), now ASTM C1260–94 [58]

- has been used for nearly a decade and has shown exceptionally good correlation between field experience and the results of the test. The test involves making mortar bars to the specification given in ASTM C227 [59], immersing the bars in water first and then molar caustic soda solution, and measuring the change in length of the bars over 12 days. Most stable aggregates show expansions of around 0.05% in this period while highly expansive ones can show expansions of more than ten times this figure. There are a number of variants of this type of test which use differing sizes of mortar bars and differing conditions of storage.

Rapid chemical test (ASTM C289 [60], BS 812 [37])

- has been widely used but is unsatisfactory in that it may produce a very large error and is unsuitable for many siliceous aggregates. A particular difficulty is that the result obtained can depend upon the extent to which the aggregate is cleaned after crushing and it can take sometimes 20 separate washing steps in order to remove the fine dust.

Mortar bar test (ASTM C227 [59], BS 812 [37] and French [61])

- has formed the basis of the more recent mortar bar tests but is in disrepute [1] because it sometimes provides unreliable data. For example, Ranc, Henri, Clement and Sorrentinop [62] have shown that six deleteriously expansive coarse aggregates were not correctly evaluated by either the ASTM C227 mortar bar test, or the ASTM C289 [60] chemical test. The ASTM C227 test also requires several months to over a year to provide indicative data.

Gel pat test (BS 812 [37], Jones and Tarleton [63])

- has been widely used. This consists of a pat made using a cement and water mixture with a water/cement ratio of 0.4 in which are embedded some 200 to 300 pieces of the aggregate in a selected size range. The flattened and polished surface of this pat is then exposed to a strongly alkaline solution and the pat surface is checked on a regular basis usually over a period of a week. In experienced hands the test is satisfactory for the detection of highly reactive materials, but there are sometimes false results arising from the occurrence of exudations on the pat surface which are not alkali–silicate gel.

unaffected by the surrounding Portland cement paste but a few exhibit significant changes which can affect the performance of the concrete. Some of these changes may occur very rapidly, while others may continue for several decades. The effects can vary from producing minor disfigurement to structural damage. In most cases under the microscope it can be seen that internal cracking occurs, contributing to the deterioration of the concrete, and although cracks may be observed in thin sections, they are better seen on polished plates or sometimes on sawn or broken surfaces.

According to French [64], the most common types of change are:

- shrinkage of particles
- softening of the aggregate
- expansion
- inorganic chemical reactions
- chemical reactions involving organic materials.

This categorization is broadly followed here.

Shrinkage of particles

Shrinkage of clay-rich particles is extremely common, although these do not usually form a large proportion of selected aggregate. The term 'clay' refers to natural material composed of particles in a specified size range, generally less than 2 μm, which forms the bulk of 'argillites', i.e. the clay-rich rocks.

Mineralogically, clay refers to a group of layered silicate minerals including the clay–micas (illites), the kaolin group, very finely divided chlorites, and the swelling clays – smectites (montmorillonites). Members of several groups, particularly micas, chlorites, and vermiculites, occur both in the clay-size range and in larger sizes. Some clays are made up of alternating layers of two or more clay groups. Random, regular, or both types of interlayering are known. If smectite is a significant constituent in such mixtures, then fairly large volume changes may occur with wetting and drying. Ferruginous argillites (i.e. mudstones) such as ironstones are generally more abundant and may show shrinkage cracks internally.

A number of instances have been reported [64] of shrinkage of materials which are initially robust, including greywacke, shale and mudstone (drying shrinkage > 0.085%) and slightly altered rocks containing some clay minerals including some dolerites and basalts (drying shrinkage 0.066% to 0.085%). Granites, limestone, quartzite, felsite, gabbroic rocks, flint and marble can show shrinkage of less than 0.05% [20].

Softening of the aggregate

Softening can take place when the aggregate includes weak or porous particles. Most aggregate particles composed of or containing large proportions of clay minerals are soft and, because of their large internal surface area, porous. Some of these aggregates will disintegrate when wetted. Rocks in which the cementing matrix is principally clay (e.g.

clay-bonded sandstones) and rocks in which swelling clays (e.g. smectite) are present as a continuous phase or matrix (e.g. some altered volcanics) may slake in water or disintegrate in the concrete mixer. Rocks of this type are unsuitable for use as aggregates. Rocks with these properties less well developed will abrade considerably during mixing, releasing clay, and raising the water requirement of the concrete containing them (Fig. 6.6). When such rocks are present in hardened concrete, the concrete will manifest greater volume changes on wetting and drying than similar concrete containing non-swelling aggregate.

Under the microscope it is fairly common to see the ferruginous matrix of some fine-grained rocks dispersed into the surrounding paste with the remnant of the rock becoming extremely weak. Pyrite may become converted to a weak mixture of hydrated iron oxide.

Some types of aggregate also become softened through alkali–aggregate reaction with the generation of gel within the aggregate displacing material and leading to a general weakening.

Expansion

Alkali–silica reaction This expansive reaction, which occurs between certain types of siliceous aggregate and the alkali metal hydroxides, although a relatively recent discovery in the UK, was known to have caused damage in the USA to concrete placed as early as 1914. By 1938 a number of bridges and roads in California were found to be affected by serious cracking, in some cases within only five years of constuction, and in 1940 Stanton [65] published the results of investigations into causes.

It was found that the fine aggregates which had been used in all the affected concrete were river sands of the same general type, containing shales, cherts and siliceous limestones. It was also found that the reactive sands only caused expansion when used with cements of alkali content greater than 0.6% equivalent Na_2O. Following the publication of Stanton's paper, investigations were made throughout the USA of concrete structures exhibiting the same type of distress and of the aggregates used in their construction. The minerals responsible for expansions were identified as opal, chalcedony, tridymite and cristobalite. Volcanic rocks of acid to intermediate composition, including rhyolites, dacites and tuffs, were also found to cause expansion. The reactive minerals contain silica in non-crystalline or crypto-crystalline forms and differ in reactivity according to the degree of disorder in their crystal structure. Opal, which is amorphous or highly disordered, is the most reactive. Chalcedony, tridymite, cristobalite, the glasses, and the crypto-crystalline and micro-crystalline forms of silica are less so. Geologically strained quartz (e.g. by heat or pressure), which can occur in quartzite rocks and the sands derived from them, has also been found to be reactive, but quartz having a well-ordered crystal structure is unreactive [66].

The first well-published case history of alkali–silica reaction in the UK was recognized in 1971 in a dam on Jersey and was attributed to the presence of opal and chalcedony in the diorite crushed rock and beach sand aggregates used [67]. The reaction was later found to have occurred in concrete in south-west England, South Wales and the Trent Valley of the English Midlands, and elsewhere to a lesser extent. It was at first suspected that marine aggregates might have been responsible since reactive flint sands and gravels dredged from the North Sea were known to have caused considerable damage in Denmark and northern Germany. It was found, however, that only about half of the affected concrete had been made using sea-dredged material [68].

The relatively late discovery of alkali–silica problems is not confined to the UK. For example, as late as 1982 the Japanese did not consider that they had a significant problem with AAR. By 1986 the problem had become important enough for the Japanese to send a large delegation to the Seventh International Conference on AAR in Ottawa and the Eighth International Conference held in Japan. In 1991 the Norwegian delegation to the International Conference on Large Dams, in Vienna [69], advised that the problem had only just been identified in Norway with 20 dams diagnosed as being affected.

The chemical processes involved in the alkali–silica reaction are fairly straightforward. The alkali metal hydroxides present in the pore water of the cement paste attack the reactive forms of silica to give a gel of alkali silicates of variable composition. Although calcium hydroxide does not take part in the attack, it can enter the alkali–silicate gel from the pore water, and form alkali–calcium silicates. The reaction products need to absorb water to swell and to grow in volume, and this causes a swelling pressure to develop within the concrete. As more water is absorbed the gel becomes more fluid and able to flow into cracks and voids in the concrete and it may reach the surface and appear as wet spots or as exudations.

The factors which determine whether gel formation and swelling will result in sufficient expansion to crack the concrete are complex. There must be sufficient alkali hydroxide present to react with the silica and enough water to maintain a minimum moisture content of about 75%

within the concrete. The swelling is also influenced by the composition of the gels, those consisting essentially of alkali silicates being more expansive than those containing a high proportion of calcium [70]. The formation of gel is dependent on the presence of reactive silica in the aggregate but this need be present in only a minor amount. This critical content of reactive material is termed the 'pessimum' value of an aggregate or aggregate combination. With highly reactive forms of silica, for example opal, a few per cent is sufficient to cause maximum expansion; with slowly reacting aggregates the pessimum proportion is higher, and in some cases, maximum expansion may not occur unless the aggregate consists entirely of reactive material. The size of the aggregate particles which contain reactive silica is a factor in the expansion, since there must be sufficient reactive material at any point within the concrete to produce more gel than can be accommodated in the neighbouring pore space. Very fine particles of known reactive materials may not cause expansion. Indeed, Stanton found that an opaline chert, which caused damage when used in normal aggregate gradings, produced little expansion when finely crushed and used as a 15–20% cement replacement. In the damaged concretes examined in England, the reactive silica was found mainly in the 1–5 mm particle size [68].

Where reaction centres are in sufficiently close juxtaposition, the cracks tend to join and cause general expansion of the concrete. For this to occur, the gel produced must be sufficiently viscous not to leak into the abundant pore space available within most concrete [1].

Because of the complexity of the factors which produce expansion it is difficult to be categorical regarding aggregate types and combinations which might be susceptible. Rock types unlikely to be reactive are given [71] as those belonging to the granite, gabbro, basalt (except for andesite), limestone (except for dolomites), porphyry (except for dacites, rhyolites and felsites) and hornfels group classifications of BS 812 [37], but concrete made with innocuous coarse aggregate may still expand if the fine aggregate (e.g. natural sand) contains reactive material. This has been found in some of the affected concrete in the UK where the coarse aggregate was limestone or granite and expansion was caused by the reactivity of the fine flint aggregate used. For further reading on recent developments, see Wood [72] and French [1].

It is probably prudent to agree at this stage with French [1], who concludes that as the list of reactive minerals given in the literature is so long, it is safest to regard all aggregate sources as possibly containing rock types that can take part in adverse alkali reactions. The magnitude of reaction of the many potentially reactive rock types varies from trivial and insignificant to highly significant, which he conveniently divides into three grades as follows.

1. Gel is produced and perhaps a little microcracking occurs but no intersecting fine cracks are present. This level of reaction is widespread and insignificant and requires a careful search for its identification.
2. Gel and intersecting fine cracks are produced internally and some slight surface expression of cracking is encountered. This is the most commonly occurring condition of reaction. In many cases it may not be obvious on a superficial examination of the structure.
3. Cracking occurs which leads to spalling and obvious surface effects and may lead to localized disintegration.

Alkali–carbonate reaction The mechanism of alkali–carbonate rock reaction is complex and disputed. It is agreed that the first stage is attack by the alkali metal hydroxides on dolomite crystals, giving calcite, brucite (hydrated magnesium oxide) and alkali carbonates – a process which has been termed dedolomitization because, in effect, it destroys the dolomite in the rock.

There is no general agreement on whether this chemical reaction is directly responsible for expansion or whether it has an indirect effect in that it provides conditions favourable for other reactions to occur. Gillot and Swenson [73], on Canadian evidence, advocate the indirect mechanism and argue that if the chemical reaction alone were responsible all dolomitic rocks would cause expansion when used as concrete aggregates, and consider that the prime cause of expansion is due to clay, which is present in the reactive rocks. They propose the theory that the water which was held in the lattices of the clay minerals at the time the sedimentary deposits were laid down was later expelled by consolidation and other rock-forming processes. After lithification of the deposits, water was unable to penetrate into the fine-grained rocks to rewet the clay minerals, unless channels for its entry were opened by chemical or physical action. In concrete, it is suggested that microcracks resulting from the dedolomitization of the aggregate provide the necessary channels. On rehydration, the desiccated clay minerals increase in volume and the swelling pressures thus created cause the concrete to crack. If this explanation is accepted, bearing in mind that in Canada reactive rocks were only found at considerable depths in the quarries, it is difficult to see how alkali–carbonate reaction can cause damage to concrete made with dolomite or dolomitic limestone rocks of evaporite origin such as occur in the Middle East [66].

Reaction rims often form around aggregate particles in concrete made with crushed carbonate rocks, but studies have shown that they do not give a positive indication of alkali–carbonate reaction having taken place. Aggregates developing rims were generally found to be similar in composition to the reactive rocks, but not all proved to be expansive.

Conversely, some rocks which caused expansion did not develop reaction rims [74, 66].

Chemical reactions with sulphates

The presence in aggregates of any type of soluble sulphate in other than limited amounts can cause chemical attack to take place in Portland cement concrete.

Carbonate rocks and shales may contain sulphates as impurities. The most abundant sulphate mineral is gypsum (hydrous calcium sulphate, $CaSO_4 \cdot 2H_2O$); anhydrite (anhydrous calcium sulphate, $CaSO_4$) is less common. Gypsum is usually white or colourless and characterized by a perfect cleavage along one plane and by its softness, representing hardness of 2 on the Mohs' scale; it is readily scratched by the fingernail. Gypsum may form a whitish or crystalline coating on sand and gravel. It is slightly soluble in water.

Anhydrite resembles dolomite in hand-sized specimens but has three cleavages at right angles; it is less soluble in hydrochloric acid than dolomite, does not effervesce and is slightly soluble in water. Anhydrite is harder than gypsum.

Sulphates react with the hydrated aluminates in the cement to give high-sulphate tri-calcium sulphoaluminate (or ettringite). This occupies more than twice the molecular volume of the aluminate and its formation in hardened pastes is accompanied by expansive forces which can exceed the tensile strength of the concrete. Sulphate attack is more rapid and severe with cements of high tri-calcium aluminate content, but even sulphate-resisting Portland cements are not immune to the effects of large amounts of sulphate in the concrete.

Needle-like (acicular) phases can be produced as a result of reaction between sulphates (gypsum or anhydrite). Although any finely divided gypsum in the cement paste reacts during the early hydration, the larger crystals of gypsum will become progressively replaced by calcium hydroxide crystals. The concomitant liberation of sulphate ions into the surrounding paste leads to the formation of ettringite in the paste and in microcracks and can be extremely damaging. The most damaging circumstances are often when concrete containing crystals of gypsum is maintained dry for a considerable period and then becomes wet; expansion with abundant cracking can then occur extremely rapidly. Similar very rapid expansion would be expected from the presence of altered sulphides which become wetted [64].

Damage to concrete caused by sulphates in natural aggregates is not an extensive problem in the UK, where the sulphates occur chiefly in clay strata. Pit sands and gravels and crushed rock aggregates are naturally of low sulphate content.

Barium sulphate, barytes, and strontium sulphate (celestite), are of such low solubility in water that they do not contribute to sulphate attack. Baryte is used as a heavy aggregate in concrete for nuclear shielding. Celestite, while of rarer occurrence than gypsum, is known to be present in small amounts in some aggregates, notably in the Middle East. Its solubility in acids is also low, and it goes undetected in the usual wet chemical tests for sulphate content. Deposits of the very soluble sulphates of magnesium (Epsom salt) and of sodium (Glauber's salt) occur naturally but usually not in rocks which yield usable aggregates [66]. In regions of the world where surface temperatures can exceed 40°C, the gypsum may be partially or completely dehydrated through the bassanite series to sulphates containing less than two molecules of water to anhydrite. Sulphates (and other evaporite salts) are common components of sabkhas and salinas in hot arid lands, and contamination for example, from desert sands, is a constant risk [75].

Because of its high sulphate content, 2 g/l as SO_3, it might be thought that residual sea water would be a source of contamination of marine-dredged sands and gravels, but that is not the case. The sea-won materials around the UK are composed principally of flint, which has a low water absorption, so any contamination is on the surface only. If it is assumed that aggregate stockpiles retain, say, 10% of water after draining, even unwashed marine aggregates would not contain more than 0.02% of sulphate.

Chemical reaction with sulphides

Damaging chemical reactions can occur with rock types containing sulphides, often leading to the formation of iron oxides and sulphate ions which combine with the calcium-rich material of the Portland cement paste to generate gypsum. These processes commonly lead to surface spalling and staining of concrete surfaces, particularly where the sulphide is disseminated as very fine particles in rocks such as argillites and meta-argillites. Any occurrence of certain types of pyrite, pyrrhotite or marcasite is likely to produce this surface damage.

The sulphides of iron, pyrite, marcasite and pyrrhotite, are frequently found in natural aggregates. Pyrite (FeS_2) is found in igneous (especially gabbros), sedimentary and metamorphic rocks (especially metamorphic limestones); marcasite is much less common and is found mainly in sedimentary rocks; pyrrhotite is less common but may be found in many types of pegmatites, basic igneous and metamorphic rocks. Pyrite is brass-yellow, and pyrrhotite bronze-brown, and both have a metallic lustre. Pyrite is often found in cubic crystals. Marcasite is also metallic but lighter in colour; finely divided iron sulphides are black. Marcasite readily oxidizes with the liberation of sulphuric acid and formation of iron oxides, hydroxides and, to a much smaller extent, sulphates; pyrite and pyrrhotite do so less readily.

If the concrete is extremely wet and sufficiently porous for the environment to be oxidizing, the sulphides may liberate sulphate ions deeper in the material. These ions react with aluminous phases to produce the solid solution series that includes ettringite. Expansion and cracking may result.

The chemical reactions do not inevitably occur when pyrite is present in concrete aggregate. Midgley [76] investigated the staining of concrete made with Thames river gravels, containing pyrite, and recognized two forms, reactive and unreactive, distinguished by their behaviour in saturated lime water. A few minutes after immersion the reactive varieties produce a green-blue precipitate of ferrous hydroxide which rapidly changes to a brown precipitate of ferric hydroxide. He found no relation between the physical forms of pyrite and its reactivity, but chemical and spectrographic analyses showed that the unreactive forms contained more additional metals (antimony, arsenic, copper, lead, manganese, tin, zinc) than did the reactive varieties. The analyses also showed that both forms contained less sulphur than the two atoms denoted by the chemical formula. This is not unusual since pyrite has a defective structure and rarely contains stoichiometric proportions of iron and sulphur, the latter usually being deficient. Midgley concluded that the extra metals in the unreactive pyrite in some way stabilized the structure so that the reaction with lime water did not occur.

Chemical reaction with chlorides

Chlorides in very small amounts are usually present as contaminants (but not necessarily getting into the concrete with the aggregates) and do not react expansively with Portland cements as do sulphates. They can be tolerated in plain concrete, although when present in large amounts some surface dampness may result. Their effect when present is to increase the risk of corrosion of embedded metals in concrete, and widespread and serious damage has been caused by the use of chloride-contaminated aggregates in reinforced concrete. Particularly at risk are thin pre-stressing wires. The corrosion products occupy more than twice the volume of the steel and their formation can be accompanied by pressures as great as 32 N/mm^2 [77, 78], resulting in cracking of the concrete, frequently followed by spalling of the cover. In severe cases of corrosion there may be a reduction in section of the reinforcing bars, leading to a loss of tensile strength in the concrete structure. A large and growing publication list exists on this topic.

The critical chloride content, indicating incipient danger of corrosion, depends on various parameters. There is therefore no single generally valid value of critical chloride content. The situation is shown in Fig. 6.13. If the concrete is not carbonated, 0.05% chloride by weight of concrete, or 0.4% chloride by weight of cement, is a good criterion for incipient danger of corrosion. However, as Fig. 6.13 shows, the critical value can be much higher or lower, depending on the environmental influences [21].

Since prestressing steels are more sensitive to corrosion, a lower limit of about 0.025% chloride by weight of concrete, or 0.2% chloride by weight of cement, is often recommended for prestressed structures.

Chlorides must not be deliberately added to the concrete mix, regardless of whether or not an agreed maximum chloride content would be exceeded. During the hydration of Portland cements, part of

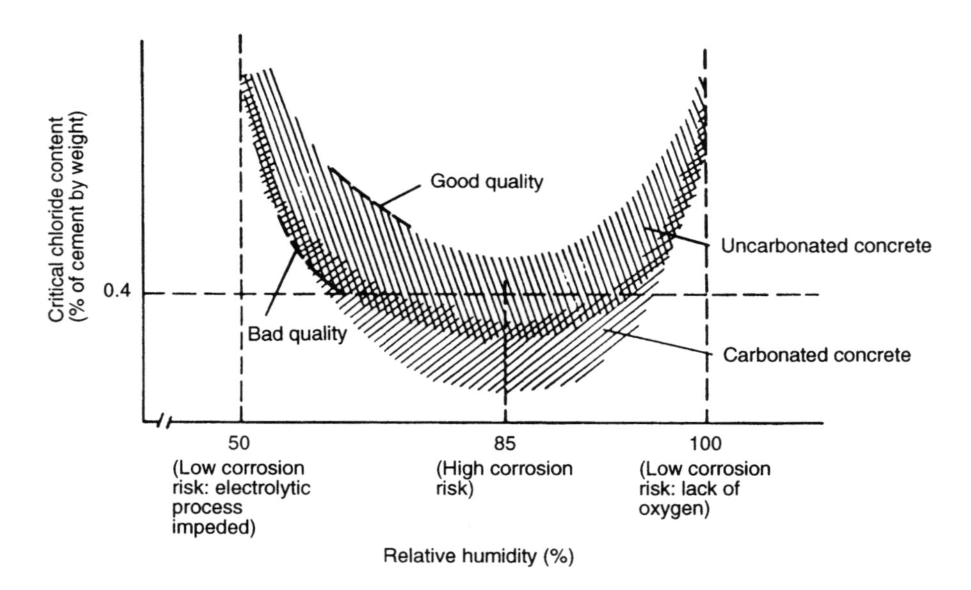

Figure 6.13 Variation of critical content with environment. (Reproduced from Comité Euro-International du Béton, *Durable Concrete Structures: Design Guide*, 2nd edn; Thomas Telford, 1992.)

any chloride present reacts with the aluminates and ferrites to give the solid compounds tri-calcium chloroaluminate and tetra-calcium chloroferrite. It has been stated that perhaps 90% of the chloride can be combined in this way with ordinary Portland cement [79]. The chloride which remains in the pore water is responsible for corrosion of the steel. It can be argued that in cements of low aluminate content less chloride can be combined as chloroaluminate and a greater proportion will remain in the free state. Corrosion would therefore be initiated at lower chloride levels in sulphate-resisting Portland cement concrete than in ordinary Portland cement concrete. The aluminates cannot react with unlimited amounts of chloride, since the sulphate in the cement combines preferentially and, therefore, the higher the sulphate content, the greater the amount of chloride that remains in the pore water. A potentially dangerous situation thus exists when aggregates are contaminated with both sulphates and chlorides; such a situation commonly exists in coastal areas in hot arid countries.

A further factor is involved when water is able to pass through the concrete. The solid chloroaluminates and chloroferrites are in equilibrium with the free chloride in the pore water. If any of this is removed by leaching, chloride from the solid phases will pass into solution to restore the equilibrium and hence corrosive conditions will be maintained in the concrete [80].

In the UK, although some land-won sands may contain small amounts of chlorides, the aggregates most likely to be contaminated are sea-dredged materials. Since, on average, sea water contains 1.8% of chloride as chlorine (higher in warmer parts of the world and shallow, enclosed areas), the possibility exists for significant amounts to be left on the aggregates unless adequately washed. This was a matter of some concern when marine aggregates first became widely used in the UK, and in 1968 the Greater London Council [81] issued a specification governing the use of marine aggregates in their works. However, more experience in the use of washed sea-dredged materials in the UK has now shown that only exceptionally do they have chloride contents exceeding the limits given in standard specifications.

In arid countries, especially coastal areas, the available aggregates (both sands and crushed carbonate rocks) may be heavily contaminated with evaporites, especially chlorides. Sodium chloride is most prevalent but minor amounts of magnesium and potassium chlorides may also be present [75].

Chemical reactions with other inorganic minerals

Other chemical effects on aggregate include dedolomitization of dolomitic limestones. The release of magnesium into the Portland cement paste can lead to the formation of brucite or hydrated magnesium silicates. Both of these phases have little or no cementitious value, i.e. a weakening of the concrete matrix could occur. There may also be an overall expansion in this chemical exchange, but the dedolomitization is not always expansive [64] (see above).

It is also theoretically possible that olivine in contact with an alkaline environment, such as cement paste, could be converted into serpentine minerals. Whether or not this alteration is expansive, however, depends on the concomitant changes occurring in the surrounding paste. Experiments have shown that this process is not significant at normal temperatures although in the event of such concrete coming into contact with a damp environment at an elevated temperature, of perhaps 200°C, some expansive reaction might occur [82].

Chemical reactions involving organic materials

Certain types of organic matter, if present in significant amounts in aggregates, can retard the rate of setting and hardening of Portland cements. The active substances are present in the humic compounds derived from the decay of plant and animal tissues, and are most often associated with silt and clay-sized material in pit sands. A distinction must be made between the presence of recognizable particles of incompletely decomposed plant material such as lignin, whose effect, if any, is to cause staining of concrete surfaces, and the true humic colloids containing the chemical compounds which retard the setting of Portland cement.

> Soil organic matter which may on rare occasions contaminate aggregate, is chemically very complex, since it is subject to continual processes of microbial degradation and synthesis. During these processes readily assimilable and therefore relatively short-lived compounds are formed. These have definite physical and chemical properties and include sugars, proteins, fats, waxes and low molecular weight organic acids. The residual humic substances, which are intermediates in the conversion of plant residues into carbon dioxide and inorganic salts, have no specific chemical or physical properties, but consist of fairly stable, dark-coloured, amorphous material, which has been estimated to have a mean residence in the soil of 1000 years [83].

An organic chemical exchange reaction sometimes encountered is the removal of bituminous materials from some limestones. It is normally extremely difficult to dissolve or remove the bituminous material that occurs in many limestones. However, the alkaline pore fluid of concrete appears to be able to extract much of the bituminous material from limestone so that the surrounding paste becomes stained. The presence of organics of this type, if abundant, could have effects on the setting rate of the paste [64].

A more obvious and common effect is that bitumen can be removed from the aggregate and deposited on the surface of the concrete, resulting in staining. This may be a general mottling with brownish patches alternating with more normally coloured concrete or the development of streaks and tear drops of dark brown material arising from individual aggregate particles. Experiments have confirmed the release of bitumen from limestone (Fig. 6.14) by alkalis in the pore fluid of the cement paste [64]. The process has been shown to cause alkalis to migrate through the porous, bituminous limestones and to become concentrated on the surface of concrete where the aggregate is close to the concrete surface. In addition, the experiments have shown that halos of bituminous material develop around the bituminous aggregate throughout the concrete.

6.5 CONCLUSIONS

Quantification of the performance characteristics of aggregates in concrete is still in its relative infancy and, at present, rigorous models do not exist for such prediction. Although not discussed in this chapter, deterioration related to chloride penetration (mainly from the external environment) is being developed in a more rigorous fashion and predictive models have been established. This is also true, but to a lesser extent, for alkali–silica reaction, where predictive models are being developed related to the alkali content of the concrete (or the ability of the concrete to acquire alkali ions). Again this has not been discussed in this chapter.

In the author's view, the best approach for the present remains the observation and study of the performance of aggregates in real concretes in different environmental and structural situations. Deterioration processes are so complex with the interaction of the environment, exposure, workmanship, design and other factors (e.g. carbonation or chloride ingress), in addition to the performance of the aggregate, that a rigorous predictive model approach is not yet realistic.

It follows, therefore, that the best way to prevent problems in concrete developing from aggregate deterioration is to carry out a thorough field and laboratory testing appraisal of the aggregates intended for use. The study of the quarry is as important as, or perhaps more important than, the study of the aggregates, because this will enable a prediction to be

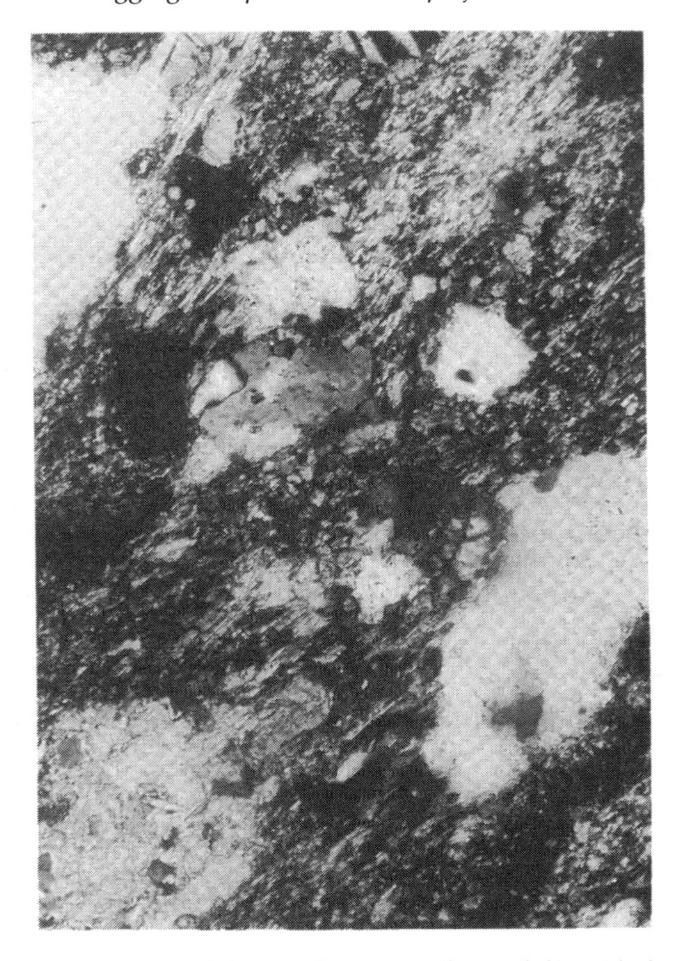

Figure 6.14 Bituminous dolomitic limestone (lower left) with bitumen in surrounding paste due to alkali leaching. Width of photomicrograph 1 mm. (Reproduced with permission from Dr W.J. French.)

made of variations of potential deleterious materials in the quarry which might otherwise get into the aggregate undetected.

Potentially dangerous situations exist when a list of permissible aggregate rocks is specified, because this does not take into account the possibility of there being unspecified deleterious material present in the rock mass of the quarry.

A list of rocks that have been found to contain deleterious material in the past is perhaps a better approach, although again this should be viewed with considerable caution. In different geological situations the same rock may be free from deleterious material, or it may be somewhat genetically different or altered in a slightly different manner from normal, making it perfectly adequate as aggregate even though banned by the specification.

ACKNOWLEDGEMENTS

Parts of this chapter have been drawn from the Fifth Sir Frederick Lea Memorial Lecture [3] presented to the Institute of Concrete Technology, April 1992. Thanks are due to my wife for wordprocessing and overcoming the intransigence of the graphics; to Dr W.J. French (Queen Mary and Westfield College, London) for photomicrographs, Figs. 6.5–6.8, 6.10–6.12, and 6.14; to Mr G. Pettifer (engineering geologist, London) for draughting Fig. 6.9, and Mr J.M. Manning (civil engineer, Winchester) for proofreading and helpful comments.

REFERENCES

1. French, W.J. (1995) Avoiding concrete aggregate problems, in *Improving Civil Engineering Structures – Old and New* (ed. W.J. French), Geotechnical Publishing Ltd, Essex, UK, pp. 65–95.
2. Smith, M.R. and Collis, L. (eds) (1993) *Sand, Gravel and Crushed Rock Aggregates for Construction Purposes*, 2nd edn. Report by Geological Society Engineering Group Working Party on Aggregates. The Geological Society, London.
3. Fookes, P.G. (1992) Geology and concrete: a review of British aggregates, The Fifth Sir Frederick Lea Memorial Lecture, in *Proceedings of Twentieth Annual Convention of Institute of Concrete Technology*, April, Coventry, UK.
4. Fookes, P.G. (1991) Geomaterials. *Quarterly Journal of Engineering Geology*, **24** (1), 3–25.
5. Building Research Establishment (1978) *The Use of Crushed Rock Aggregates in Concrete*. RR18.
6. Building Research Establishment (1986) Porous aggregates in concrete: Jurassic limestones. *Information Paper IP 2/86*.
7. Building Research Establishment (1989) Porous aggregates in concrete: sandstones from NW England. *Information Paper IP 16/89*.
8. Building Research Establishment (1987) Sea dredged aggregates in concrete. *Information Paper IP 7/87*.
9. Readymix Concrete (1986) *Marine Aggregates – Off-shore Dredging for Sand and Gravel*, RMC, Feltham, UK.
10. Cement & Concrete Association (1984) *Expansion of Concrete Due to Alkali–Silica Reaction*, Reprint 1/84, C&CA, Slough, UK.
11. British Cement Association (1992) *The Diagnosis of Alkali–Silica Reaction*. Reference 45.042, BCA, Watford, UK.
12. Property Services Agency (1979) *Standard Specifications Clauses for Airfield Pavement Works, Part 3*, PSA, Croydon, UK.
13. British Standards Institution (1981) *Ready Mixed Building Mortars*. BS 4721, BSI, London.

14. British Standards Institution (1986) *Specifications for Building Sands from Natural Sources: Sands for Mortars for Plastering and Rendering*. BS 1199, BSI, London.
15. Building Research Establishment (1980) *Materials for Concrete*. Digest 237. Her Majesty's Stationery Office, London.
16. Teychenné, D.C. (1978) *The Use of Crushed Rock Aggregates in Concrete*. Building Research Establishment, Watford, UK.
17. Mitchell, L.J. (1956) *Thermal Properties*, Special Technical Publication 169, American Society for Testing Materials, Philadelphia, pp. 129–35.
18. British Standards Institution (1992) *Specifications for Aggregates from Natural Sources for Concrete*. BS 882, BSI, London.
19. British Standards Institution (1973) *Aggregates from Natural Sources for Concrete (Including Granolithic)*. BS 1201, BSI, London.
20. Building Research Establishment (1968) *Shrinkage of Natural Aggregates in Concrete*, Digest 35. Her Majesty's Stationery Office, London.
21. Comité Euro-International du Bréton (1992) *Durable Concrete Structures: Design Guide*, 2nd edn, Thomas Telford, London.
22. Fookes, P.G. and Collis, L. (1976) Cracking and the Middle East, *Concrete*, **10** (2), 14–19.
23. Kay, T. (1990) *Assessment and Renovation of Concrete Structures*, Longman.
24. Mays, G. (ed.) (1992) *Durability of Concrete Structures: Investigation, Repair and Protection*, E. & F.N. Spon, London.
25. Rostam, S. (1991) Philosophy of assessment and repair of concrete structures, and the feedback into new designs, in *Damage Assessment Repair Techniques and Strategies for Reinforced Concrete*, (ed. G. L. MacMillan), Proceedings of the Bahrain Society of Engineers Regional Conference, December, pp. 33–126.
26. Blyth, F.G.H. and de Freitas, M.A. (1984) *A Geology for Engineers*, 7th edn, Edward Arnold, London.
27. McLean, A.C. and Gribble, C.D. (1979) *Geology for Civil Engineers*, Allen and Unwin, London.
28. British Geological Survey (1990) *United Kingdom Minerals Yearbook*, Keyworth, Nottingham, UK.
29. British Standards Institution (1981) *Code of Practice for Site Investigations*. BS 5930, BSI, London.
30. Fookes, P.G. (1980) An introduction to the influence of natural aggregates on the performance and durability of concrete. *Quarterly Journal of Engineering Geology*, **13**, 207–29.
31. Cole, R.G. and Horswill, P. (1988) Alkali–silica reaction at Val de la Mere Dam, Jersey, case history. *Proceedings of the Institution of Civil Engineers, Part 1*, **84**, 1237–59.
32. Cement & Concrete Association (1970) *Impurities in Aggregates for Concrete*. Advisory Note 18, C&CA, Slough, UK.
33. American Society for Testing Materials (1981) *Standard Specification for Concrete Aggregates*. Designation C33–81, ASTM, Philadelphia, USA.
34. The Concrete Society (1987) *Alkali–Silica Reaction, Minimizing the Risk of Damage to Concrete: Guidance Notes and Model Specification Clauses*. Technical Report (the Hawkins Committee), The Concrete Society, Slough, UK.
35. Department of Transport (1986) *Specification for Highway Works*, Her Majesty's Stationery Office, London.
36. Neville, A. (1973) *Properties of Concrete*, 2nd (metric) edn, Sir Isaac Pitman & Sons Ltd, London.

37. British Standards Institution (various dates) *Methods for Sampling and Testing of Mineral Aggregates, Sands and Fillers*, BS 812, BSI, London.
38. American Society of Testing Materials (1989) *Sieve Analysis of Fine and Coarse Aggregates*. Designation C136–84, ASTM, Philadelphia.
39. American Society of Testing Materials (1989) *Specific Gravity and Absorption of Coarse Aggregate*. Designation C127–88, ASTM, Philadelphia.
40. American Society of Testing Materials (1989) *Specific Gravity and Absorption of the Aggregate*. Designation C128–88, ASTM, Philadelphia.
41. American Society of Testing Materials (1991) *Method of Test for Sand-equivalent Value of Soils and Fine Aggregate*. Designation 2419, ASTM, Philadelphia.
42. International Society of Rock Mechanics (1985) Suggested method for determining pointload strength. *International Journal of Rock Mechanics and Mining Science: Geomechanics Abstracts*, 22 (2).
43. Franklin, J.A. (1970) Observations and tests for engineering description and mapping of rocks, in *Proceedings of the Second Congress of the International Society for Rock Mechanics*, vol. 1, paper 1–3, Belgrade.
44. Duncan, N. (1969) *Engineering Geology and Rock Mechanics*, vol. 1 and 2, International Textbook Co.
45. Hosking, J.R. and Tubey, L.W. (1969) *Research on Low Grade and Unsound Aggregates*. Report LR 293, Road Research Laboratory, Crowthorne, UK.
46. Franklin, J.A. and Chandra, A. (1972) The slake-durability test. *International Journal of Rock Mechanics and Mining Science*, **9**, 325–41.
47. American Association of State Highway and Transportation Officials (1978). *Standard Specifications for Transportation Materials, Part II: Methods of Sampling and Testing*. Standard method of test for soundness of aggregates by freezing and thawing. T 103–78, Washington, DC.
48. DIN (Deutsches Institut für Normung) (1971) *Zuschlag für Beton, Prüfung von Zuschlag mit dichtem oder porgem Gefüge*. DIN 4226, Blatt 3. Deutschen Normenausschusses, Berlin.
49. American Society of Testing Materials (1989) *Soundness of Aggregates by use of Sodium Sulphate or Magnesium Sulphate*. Designation C88–83, ASTM, Philadelphia.
50. American Society of Testing Materials (1989) *Resistance to Abrasion of Small-Size Coarse Aggregate by Use of the Los Angeles Machine*. Designation C131–89, ASTM, Philadelphia.
51. American Society of Testing Materials (1985) *Recommended Practice for Petrographic Examination of Aggregates for Concrete*. Designation C295–1985, ASTM, Philadelphia.
52. British Standards Institution (1990) *Chemical Tests* BS 1377, pt3, BSI, London.
53. Ramsay, D.M. (1965) Factors influencing aggregate impact value in rock aggregate. *Quarry Managers Journal*, 49, 129–34.
54. Figg, J.W. and Lees, T.P. (1975) Field testing the chloride content of sea-dredged aggregates. *Concrete*, September.
55. British Standards Institution (1983) *Specification for Air-Cooled Blastfurnace Slag Aggregate for Use in Construction*, BS 1047, BSI, London.
56. Hooton, R.D. (1991) *New Aggregate Alkali-Reactivity Methods*, MAT–91–14, Ministry of Transportation, Ontario.
57. Oberholster, R.E. and Davies, G. (1986) An accelerated method for testing the potential alkali-reactivity of siliceous aggregates. *Cement and Concrete Research*, **16**, 181–9.
58. American Society of Testing Materials (1994) Standard method for potential

alkali reactivity of aggregates (mortar bar method), Designation C1260–94, in *ASTM Standards Yearbook: Concrete and Aggregates* vol. 4.02, pp. 648–51.
59. American Society of Testing Materials (1990). Standard method for potential alkali reactivity of cement-aggregate combinations, mortar bar method. Designation C227–90, in *ASTM Standards Yearbook: Concrete and Aggregates*, vol. 4.02, pp. 126–30.
60. American Society of Testing Materials (1994). Standard test method for potential alkali–silica reactivity of aggregates (chemical method). Designation C289–94, in *ASTM Standards Yearbook: Concrete and Aggregates*, vol. 4.02, pp. 156–62.
61. French, W.J. (1992) Comparison of the Canadian and British Standard Concrete Prism Tests and the effect of reduced permeability on the results, in *Proceedings of Ninth International Conference on Alkali–Aggregate Reaction in Concrete*. Concrete Society publication CS 104, pp. 347–32.
62. Ranc, R., Henri, I., Clement, J.Y. and Sorrentinop, D. (1994) Limits of application of the ASTM C227 mortar bar test. Comparison with two other standards on alkali–aggregate reactivity, *American Society of Testing Materials, Cement, Concrete, Aggregate*, **16** (1) 63–71.
63. Jones, F.E. and Tarleton, R.D. (1958) Reactions between aggregates and cement, *National Building Studies, Research Paper 225*, Her Majesty's Stationery Office, London, pp. 18–21.
64. French, W.J. (1991) Concrete petrography, a review. *Quarterly Journal of Engineering Geology*, **24**, 17–48.
65. Stanton, T.E. (1940) The expansion of concrete through reaction between cement and concrete. *Proceedings of the American Society of Civil Engineers*, **66**, December, 1781–1811.
66. Eglington, M.S. (1987) *Concrete and its Chemical Behaviour*. Thomas Telford, London.
67. Gutt, W. and Nixon, P. (1979) Alkali–aggregate reaction in concrete in the UK. *Concrete*, **13**, May, 19–31.
68. Palmer, D. (1981) Alkali–aggregate reactions in Great Britain – the present positions. *Concrete*, **15**, March, 24–7.
69. Mason, P.J. (1995) Concrete at dams, in *Improving Civil Engineering Structures – Old and New*, (ed. W.J. French), Geotechnical Publishing Ltd, Essex. UK, pp. 45–64.
70. Vivian, H.E. (1985) Alkali–aggregate reaction, in *Symposium on Alkali Aggregate Reaction – Preventative Measures*, August, Reykjavik.
71. Building Research Establishment (1982) *Alkali Aggregate Reactions in Concrete*, Digest 258. Her Majesty's Stationery Office, London.
72. Wood, J.G.M. (1995) Towards quantified durability design for concrete, in *Improving Civil Engineering Structures – Old and New* (ed. W.J. French) Geotechnical Publishing Ltd, Essex, UK pp. 139–159.
73. Gillot, J.E. and Swenson, E.G. (1969) Mechanism of the alkali carbonate rock reaction. *Quarterly Journal of Engineering Geology*, **2**, October, 7–20.
74. French, W.J. and Poole, A.B. (1976) Alkali aggressive aggregates and the Middle East. *Concrete*, **10** (1).
75. Fookes, P.G. (1995) Concrete in hot dry salty environments. *Concrete*, **29** (1), 34–39.
76. Midgley, H.G. (1958) The staining of concrete by pyrite. *Magazine of Concrete Research*, **10**, August, 75–8.
77. Cement and Concrete Association (1970) Impurities in Aggregates for Concrete. *Advisory Note 18*, C&CA, Slough, UK.

78. Rosa, E.B., McCollom, B. and Peters, O.S. (1913) Electrolysis in concrete *Technical Paper 18,* US Bureau of Standards.
79. Building Research Establishment (1982) *The Durability of Steel in Concrete, Part 1: Mechanism of Protection and Corrosion*. Digest 263. Her Majesty's Stationery Office, London.
80. Roberts, M.H. (1962) The effect of calcium chloride on the durability of pre-tensioned wire in pre-stressed concrete. *Magazine of Concrete Research,* **14**, November, 143–54.
81. Greater London Council (1968) Marine aggregates specification. *Bulletin 16*. GLC, London.
82. French W.J. and Crammond, N.J. (1980) The influence of serpentinite and other rocks on the stability of concretes in the Middle East. *Quarterly Journal of Engineering Geology,* **13**, 225–280.
83. Schnitzer, M. and Kahn, S.U. (1972) *Humic Substances in the Environment,* Dekker, New York.

Table 6.A.1 Characteristics of clastic sedimentary rocks

Rock type	*Geological characteristics*	*Typical[a] occurrence 1 = rare to 5 = common*	*Typical[a] quality of aggregate 1 = poor to 5 = good*	*Comments (as aggregates)*
		Coarse-grained (Rudites > 2 mm)		
Conglomerates				
General	Rounded fragments of any rock type, but quartz predominates. Cementing agent chiefly silica, but iron oxide, clay and calcareous material also common	(2)	(2 to 5)	Generally good but depends on nature of the cement matrix (e.g. could be weak)
Pudding stone	Gap-graded mixtures of large particles in a fine matrix	1	4	As General, see above
Basal	First member of a series; deposited unconformably	2	4	As General, see above
Breccia	Angular fragments of any rock type	4	2 to 4	As General, see above

Medium-grained (Arenites > 50% of grains between 0.06 and 2 mm)				
Sandstones[b]	Predominantly quartz grains cemented by silica, iron oxide, clay, or a carbonate (e.g. calcite). Colour depends on cementing agent; yellow, brown or red – iron oxides may predominate; lighter sandstones – silica or carbonate material may predominate. Porous and pervious with porosity ranging from 5% to 20% or greater. Generally strong, and thick beds common	5	3 to 5	Generally good, but types can be weak and/or porous
Arkose[b]	Similar to sandstone but with at least 25% feldspar grains	2 to 3	4 to 5	As Sandstones, see above
Greywacke[c]	Angular particles of a variety of minerals in addition to quartz and feldspar, in a clay matrix. Grey in colour, a strongly indurated, impure sandstone	2	4 to 5	As Sandstones, see above. May be weathered toward top of quarry and be of lower strength and possibly containing clay
Fine-grained (Lutites or Siltites < 2 mm)				
Siltstones[b]	Composition similar to sandstone but at least 50% of grains are between 0.002 and 0.06 mm. Seldom forms thick beds, but is often strong	3 to 4	4 to 5	Generally good. Thin bedded types may have flaky shape

Table 6.A.1 *(continued)*

Rock type	*Geological characteristics*	*Typical[a] occurrence 1 = rare to 5 = common*	*Typical[a] quality of aggregate 1 = poor to 5 = good*	*Comments (as aggregates)*
Shales				
General	Predominant particle size < 0.002 mm with a well-defined fissile fabric. Red shales are coloured by iron oxides and grey to black shales are often coloured by carbonaceous material. Commonly interbedded with sandstones and relatively soft. Many varieties exist	(5)	(1 to 2)	Generally not good – often somewhat weathered and weak. May well have flaky shape
Argillites	Strong, indurated shales devoid of fissility; and without slaty cleavage	3	1 to 3	Generally good – but may be somewhat weathered and weak. Could have flaky shape
Calcareous shales	Contain carbonates, especially calcite. With increase in calcareous content becomes shaly limestone	2	1 to 2	As General, see above. May also be porous
Oil shales	Contain carbonaceous matter that yields oil upon destructive distillation	1	1	As General, see above

Marine shales	May contain smectite clays that are subject to very large volume changes upon wetting or drying	3 to 4	1 to 2	As General, see above
Clay shales	Moderately indurated shales	3	2	As Argillites, see above
Claystones	Clay-sized particles compacted into rock	3	1 to 3	As Argillites, see above

[a] Estimated, must be considered approximate. UK rocks only.
[b] When fresh or only slightly weathered, these rocks may make suitable wearing surface concrete.
[c] Usually very good for wearing surfaces.

Table 6.A.2 Characteristics of some common non-clastic sedimentary rocks

Rock type	*Geological characteristics*	*Typical[a] occurrence 1 = rare to 5 = common*	*Typical[a] quality of aggregate 1 = poor to 5 = good*	*Comments (as aggregates)*
	Carbonate precipitates			
Limestone				
General	Contains more than 50% calcium carbonate (calcite); the remainder consists of impurities such as clay, quartz, iron oxide or other minerals. The calcite can be precipitated chemically or organically, or it may be detrital in origin. There are many varieties; all effervesce in HCl. Bulk density about 2710 kg/m^3	(5)	(2 to 5)	Generally good – check strength, abrasion and porosity of any unknown source
Crystalline limestone	Relatively pure, coarse to medium texture, strong	3	5	As General, see above
Micrite	Microcrystalline form, conchoidal fracture, pure, strong	2	5	As General, see above
Oolitic limestone	Composed of sand-sized spheres (oolites), usually contains a sand grain as a nucleus around which coats of carbonate are deposited	4	1 to 2	Generally not good, often weak and porous

Fossiliferous limestone	Parts of invertebrate organisms such as molluscs, crinoids and corals cemented with calcium carbonate	3	2 to 4	As General, see above
Coquina	Weak porous rock consisting of lightly cemented shells and shell fragments	1	1 to 2 as gravel	Generally not good, may be workable as a gravel
Chalk	Soft, weak, porous and fine textured; composed of shells of microscopic organisms; normally white colour	5	1 strong beds only	Generally not good, too weak and porous. Stronger beds may provide low-quality limestone
Dolomite	Magnesium carbonates. Stronger, harder and heavier than limestone (bulk density about 2870 kg/m^3). Forms either from direct precipitation from sea water or from the alteration of limestone by 'dolomitization'. Effervesces in HCl only when powdered. Hardness > 5	1	4 to 5	As General, see above. In new sources check for alkali–dolomite reaction

Table 6.A.2 *(continued)*

Rock type	*Geological characteristics*	*Typical[a] occurrence 1 = rare to 5 = common*	*Typical[a] quality of aggregate 1 = poor to 5 = good*	*Comments (as aggregates)*
	Biogenic and chemical origin (siliceous rocks)			
Chert	Formed of silica deposited from solution in water both by evaporation and the activity of living organisms, and possibly by chemical reactions. Can occur as small nodules or as relatively thick beds of wide extent and is sometimes common to many limestones and chalk formations, and as the limestone is removed by weathering, the chert beds remain prominent and unchanged, often covering the surface as numerous fragments. Flint is a variety of chert occurring in chalk; jasper is a red or reddish-brown chert. Hardness 7	1	4 to 5	In UK beds not usually thick enough to work as a quarry, but as a gravel (e.g. Thames ballast) often good. Check new sources for alkali–silica reaction. May be abrasive

[a] Estimated, must be considered approximate. UK rocks only.

Table 6.A.3 Characteristics of some igneous rocks

Rock type	*Geological characteristics*	*Typical[a] occurrence 1 = rare to 5 = common*	*Typical[a] quality of aggregate 1 = poor to 5 = good*	*Comments (as aggregates)*
	Coarse to medium-grained (very slow to slow cooling), mainly plutonic			
Pegmatite	Abundant as dykes in granite masses and other large bodies. Chiefly quartz and feldspar appearing separately as large grains ranging from few millimetres to as large as a metre in diameter	2	3 to 5	Usually part of a granite mass, some hydrothermally altered (as Granite, see below)
Granite[b]	The most common and widely occurring igneous rock. Fabric normally roughly equigranular. Light colours contain chiefly quartz and feldspar; grey shales contain biotite mica or hornblende	4	5	Generally no particular problems. Rock may be weathered or hydrothermally altered, especially in south-west England where it can be deeply weathered, and may contain clays. Check for inclusions of other rock types

Table 6.A.3 *(continued)*

Rock type	*Geological characteristics*	*Typical*[a] *occurrence 1 = rare to 5 = common*	*Typical*[a] *quality of aggregate 1 = poor to 5 = good*	*Comments (as aggregates)*
Syenite[b]	Light-coloured rock differing from granite in that it contains no quartz, consisting almost entirely of feldspar but often containing some hornblende, biotite and pyroxene	2	5	Generally no particular problems
Diorite[b]	Grey to dark-grey or greenish, composed of plagioclase fieldspar and one or more of the ferromagnesian minerals. Equigranular fabric	3	2 to 5	Generally no particular problems except if weathered (particularly near the ground surface in quarries) when rock will be significantly weaker and likely to contain swelling clay minerals
Gabbro[b]	Dark-coloured rock composed almost solely of ferromagnesian minerals and plagioclase feldspar	3	2 to 5	As Diorite, see above
Dolerite[b]	Dark-coloured rock intermediate in grain size between gabbro and basalt. Abundant as thick lava flows that have cooled slowly	3 to 4	4	As Diorite, see above

Peridotite	Dark-coloured rocks composed almost solely of ferromagnesian minerals. Olivine predominant; negligible feldspar. Readily altered to secondary minerals	1	3	As Diorite, see above
Dunite	Major constituent is olivine, which alters readily to serpentinite	2	2 to 3	As Diorite, see above.
Fine-grained (rapid cooling) extrusive (volcanic)				
Rhyolite	The microcrystalline equivalent of granite formed at or near the Earth's surface. Characteristically white, grey or pink and nearly always containing a few phenocrysts of quartz or feldspar	2 to 3	3 to 5	As Andesite, see below
Felsite	Occurs as dykes, sills and lava flows. The term felsite is used to define the finely crystalline varieties of quartz-porphyries or other light-coloured porphyries that have few or no phenocrysts and give little indication to the unaided eye of their actual mineral composition	2	3 to 5	As Andesite, see below

Table 6.A.3 *(continued)*

Rock type	*Geological characteristics*	*Typical*[a] *occurrence 1 = rare to 5 = common*	*Typical*[a] *quality of aggregate 1 = poor to 5 = good*	*Comments (as aggregates)*
Andesite[b]	Generally dark-grey, green or red. Pure andesite is relatively rare, and it is usually found with phenocrysts. Porphyritic andesite and basalt compose about 95% of all volcanic materials in the world (but much less in UK)	2 to 3	3	Generally no particular problems, but although not likely, need to check for alkali–silica reaction
Basalt	Most abundant extrusive rock; found in all parts of the world and beneath the oceans. Colours range from greyish- to greenish-black to black. Fine-grained with a dense compact structure. Often contains numerous voids (vesicular basalt)	5	2 to 4	Generally no particular problems (in UK) but may contain vesicles (bubble holes) filled with rare minerals (check for alkali–silica reaction) and be weaker than typical. If weathered – see Diorite above

[a] Estimated, must be considered approximate. UK rocks only.
[b] When fresh or only slightly weathered these rocks may make suitable wearing surface concrete.

Table 6.A.4 Characteristics of some metamorphic rocks with foliate fabric

Rock type	*Geological characteristics*	*Typical[a] occurrence 1 = rare to 5 = common*	*Typical[a] quality of aggregate 1 = poor to 5 = good*	*Comments (as aggregates)*
Gneiss[b]	Coarse-grained; imperfect foliation resulting from banding of different minerals. Chief minerals are quartz and feldspar, but various percentages of other minerals (mica, amphibole, and other ferromagnesians) are common. Gneiss is usually identified by its dominant accessory mineral, e.g. hornblende gneiss, biotite gneiss; or general composition, i.e. granite gneiss; paragneiss derived from sedimentary rocks; orthogneiss derived from feldspathic igneous rocks	3	3 to 5	Generally good, may be weathered towards ground surface of quarry and be a little weaker than average. Check for strained quartz–alkali reaction and inclusions of other rock types

Table 6.A.4 *(continued)*

Rock type	*Geological characteristics*	*Typical[a] occurrence 1 = rare to 5 = common*	*Typical[a] quality of aggregate 1 = poor to 5 = good*	*Comments (as aggregates)*
Schist	Fine-grained, well developed foliation, resulting from the parallel arrangement of platy minerals (i.e. schistosity). The important platy minerals are muscovite, chlorite and talc. Schists identified by the mineral as mica schist, chlorite schist, etc. Garnet is a common accessory mineral to mica schist and represents intense metamorphism. Schists and gneisses commonly grade into each other and a clear distinction between them is often not possible	3	2 to 4	As Gneiss, see above, but shape may be a little flaky. Mica or chlorite schists may be weak along foliation planes

Amphibolite	Consists largely of amphibole and shows more or less schistose form of foliation. Composed of darker minerals and, in addition to hornblende, can contain quartz, plagioclase feldspar and mica. They are hard and have bulk densities ranging from 3000 to 3400 kg/m^3. Association with gneisses and schists is common in which they form layers and masses that are often more resistant to erosion that the surrounding rocks	1 to 2	2 to 3	As Schist, see above
Phyllite	Soft, with a satin-like lustre and extremely fine schistosity. Composed chiefly of chlorite. Grades to schists as the coarseness increases	2	1	Generally not good (too weak and porous) and with flaky shape
Slate	Extremely fine-grained, exhibiting remarkable planar cleavage. Generally hard; once extensively used for roofing materials	4 to 5	1 to 2	May be suitable if not too cleaved (i.e. breaks into flaky shape)

[a] Estimated, must be considered approximate. UK rocks only.
[b] When fresh or only slightly weathered, these rocks should make good wearing surface concrete.

Table 6.A.5 Characteristics of some metamorphic rocks with massive fabric forms

Rock type	*Geological characteristics*	*Typical*[a] *occurrence 1 = rare to 5 = common*	*Typical*[a] *quality of aggregate 1 = poor to 5 = good*	*Comments (as aggregates)*
Metaconglomerate	Heat and pressure cause the pebbles in a conglomerate to stretch, deform and fuse. Strong	1	3 to 5	Generally good but can be abrasive
Quartzite	Results from sandstone so occurs across the grains, which are often imperceptible. Strong	2	3 to 5	See above
Marble	Results from metamorphism of limestone or dolomite and is found with large and small crystals, and in many colours including white, black, green and red. Metamorphosed limestone does not normally develop cavities. Strong	1 to 2	2 to 4	Generally good but may be weakened and porous if partly weathered by solution

Serpentinite	Derived from serpentine. Generally compact, dull to waxy lustre, smooth to splintery fracture, generally green in colour and often soft unless it contains significant amounts of quartz. Can have foliate fabric	1	2 to 4	Often good
Hornfels	Rocks baked by contact metamorphism into hard aphanitic material, with conchoidal fracture, dark-grey to black colour, often resembling a basalt	2	3 to 5	Generally good but may be weathered towards ground surface of quarry and be a little weaker than average
		Other forms		
Migmatite	A rock that is a complex intermixture of metamorphic and granular igneous rocks such as formed by the injection of granite	2	2 to 5	Generally good – depends on geological types, similarities with gneiss, and/or granite (see Tables 6.A.3 and 6.A.4)

Table 6.A.5 (*continued*)

Rock type	*Geological characteristics*	*Typical*[a] *occurrence 1 = rare to 5 = common*	*Typical*[a] *quality of aggregate 1 = poor to 5 = good*	*Comments (as aggregates)*
Mylonites	Produced by intense mechanical metamorphism; can show strong lamination but the original mineral constituents and fabric have been crushed and pulverized by the physical processes rather than altered chemically. Common along the base of overthrust sheets and can range from very thin, to a metre or so, to several hundreds of metres thick. They are formed by differential movement between beds	1	1	Not likely to be suitable

[a] Estimated, must be considered approximate. UK rocks only.

Table 6.A.6 Rock-mass discontinuities

Discontinuity	*Definition*	*Geological characteristics*	*Occurrence in rock*[a] *1 = rare to 5 = common*	*Significance in quarry, when occurring*[a] *1 = rare to 5 = great*	*Significance to aggregate, when occurring*[a] *0 = none to 5 = great*	*Comments*
Fabric	Total of all features of a rock (i.e. texture and structure)	The fabric controls the physical behaviour of the rock	5	5	5	See below
Fracture	A separation in the rock mass, a break	For example, joints, faults, slickensides, foliations and cleavages. Open fractures may be 'healed' with minerals (of many types, often metals)	5	5	0 to 5	See below. Healed fractures may contain physically or chemically unsound minerals
Joint	A fracture along which essentially no displacement has occurred	Most common defect encountered. Present in most rocks in some geometric pattern related to rock type and stress field. Open joints allow free movement of water, increasing irregular decomposition rate of the mass	4 to 5	4	1	Can help control aggregate shape. May also be of significance to blasting and excavation, and help control fragment shape

Table 6.A.6 *(continued)*

Discontinuity	*Definition*	*Geological characteristics*	*Occurrence in rock[a] 1 = rare to 5 = common*	*Significance in quarry, when occurring[a] 1 = rare to 5 = great*	*Significance to aggregate, when occurring[a] 0 = none to 5 = great*	*Comments*
Microjoint (or micro-fracture)	As joint but below limit of normal vision – can be seen under microscope	Common in some rock types, especially in deformed rocks (folded and faulted)	2	3	5	Can make the rock unsuitable for aggregate
Fault	A fracture caused by displacement, mainly due to tectonic activity	Fault zone usually consists of crushed and sheared rock through which water can move relatively freely, increasing weathering	1 to 3	2 to 5	1	May influence blasting and distribution of beds in the quarry
Slickensides	Pre-existing failure surface; from faulting, landslides, expansion	Shiny, polished surfaces with striations. Often the weakest elements in a mass, since strength is often near residual	1 to 2	2	3	Can help control aggregate shape and strength

Foliation planes	Continuous foliation surface results from orientation of mineral grains during metamorphism	Can be present as open joints or merely orientations without openings. Strength and deformation relate to the orientation of applied stress to the foliations	1 to 5	3	2	Will help control aggregate shape and strength
Foliation shear	Shear zone resulting from folding or stress relief	Thin zones of gouged and crushed rock occur along the weaker layers in metamorphic rocks	1 to 2	2	0	Usually removed by grizzly
Cleavage	Stress fractures from folding	Found primarily in shales and slates; usually very closely spaced	1 to 2	2	5	Can make the rock unsuitable as aggregate
Bedding planes	Contacts between sedimentary rocks	May be weak and/or contain weak materials such as lignite or smectite clays	5	4	3	Help control aggregate shape
Mylonite	Intensely sheared zone	Strong laminations; original mineral constituents and fabric crushed and pulverized	1	3	0	Usually removed by grizzly

Table 6.A.6 *(continued)*

Discontinuity	*Definition*	*Geological characteristics*	*Occurrence in rock*[a] *1 = rare to 5 = common*	*Significance in quarry, when occurring*[a] *1 = rare to 5 = great*	*Significance to aggregate, when occurring*[a] *0 = none to 5 = great*	*Comments*
Cavities, caves, karst	Openings in soluble rocks resulting from groundwater movement, or in igneous rocks from gas pockets	In limestones range from caverns to tubes. In rhyolite and other igneous rocks range from voids of various sizes to tubes	4	4	1	Little significance to aggregate except where rock is partially solution-weathered and is therefore weaker and more porous. Walls of caves may be coated with stiff mud, calcite or other material

[a] Estimated, must be considered approximate. UK rocks only.

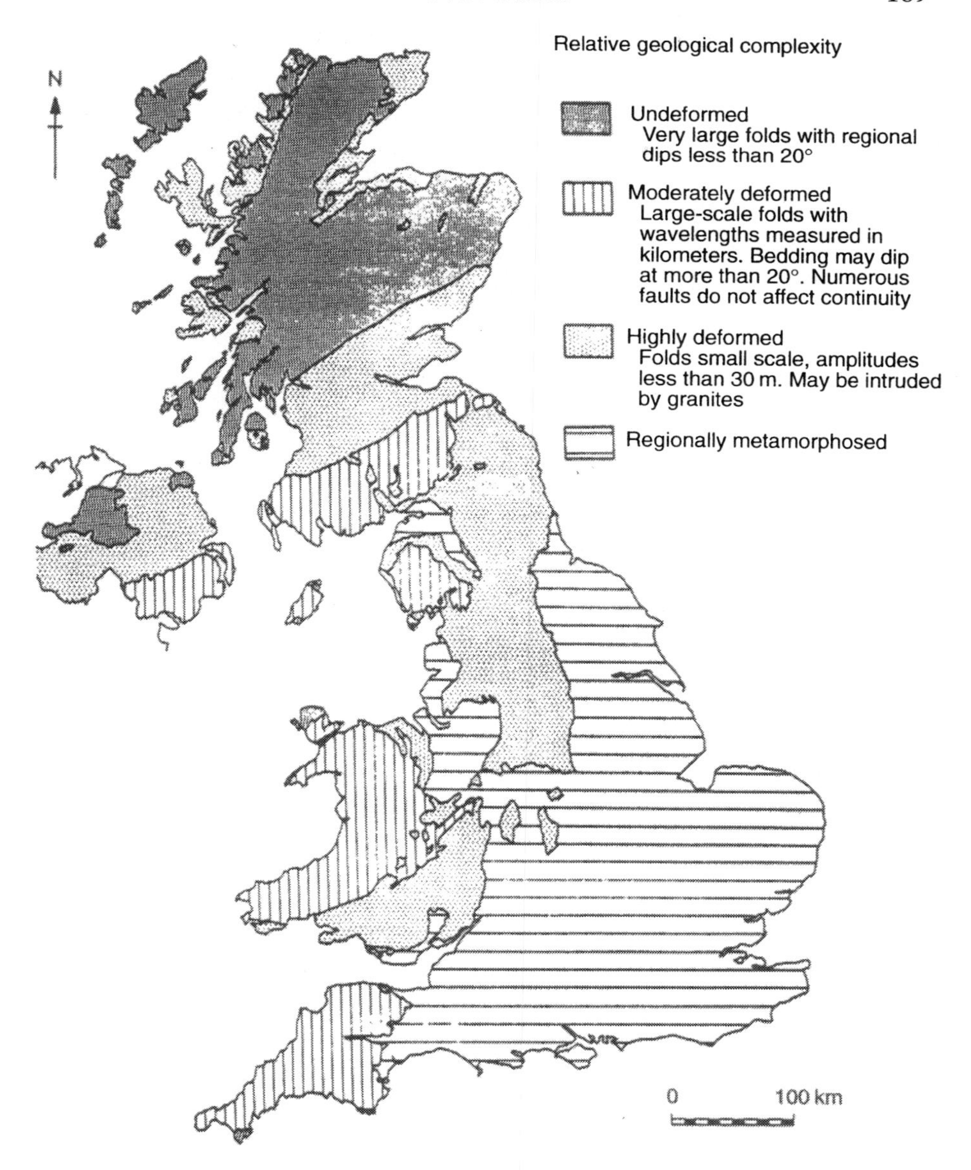

Figure 6.A.1 Distribution of the four classes of structural complexity in the bedrock. (Reproduced from Dearman, W.R. and Eyles, N., *Bulletin of the International Association of Engineering Geology*, **25**; 1982).

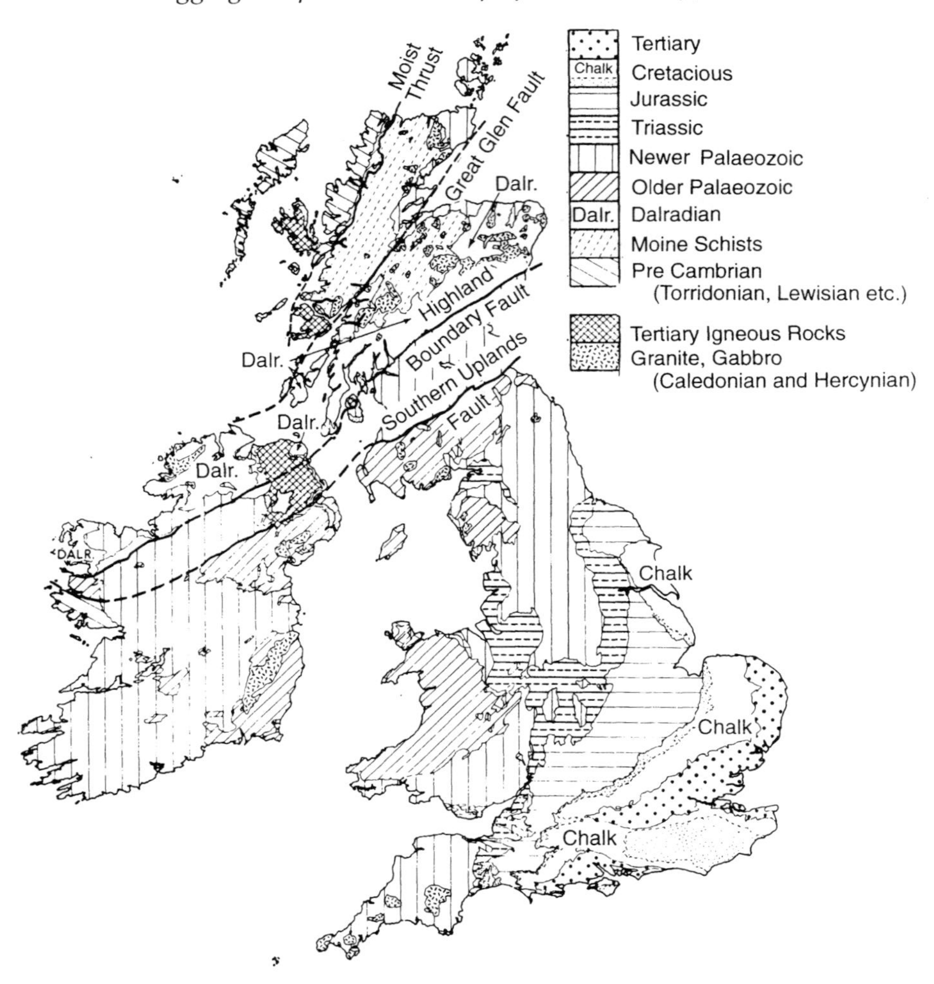

Figure 6.A.2 Outline solid geological map of UK and Ireland. (Reproduced from Blyth, F.G.H. and de Freitas, M.A., *A Geology for Engineers*, 7th edn; Edward Arnold, 1984.)

Discussion and Chairman's summing up

Dr Adam Neville We have not yet had time to discuss various papers and I think it would be best if we stay with one topic for at least five minutes before we change to something totally different. Who would like to speak first?

John Gayner, *retired contractor* I only rise because Peter Fookes has reduced the discussion to a level where a contractor can actually say something. I think it is worth remembering at the end of the day everything that has been said today has got to be turned into concrete somehow and the sad fact of the matter is that it is to be made by the people who, in theory, are paid the least in the total construction costs and have most responsibility as far as workmanship is concerned. There are reasons for everything that is done and today has underlined that workmanship is nearly as paramount as design. Professor Bungey reassures me that you can tell the strength of concrete as it actually is and not rely on cubes, because waiting 7 or 28 days before you discover you have faulty concrete is a long time and one must remember the cubes anyhow are not representative, they are in fact very specially made.

As a member of the Resident Engineer's staff at Bankside Power Station, and if anybody wishes to see 40-year old concrete I am sure the Tate Gallery will let them have a look, in the early fifties all the concrete was placed by 4 inch concrete pumps – I beg your pardon, 100 mm – and every young engineer knew in those days, and perhaps knows to this day, that if concrete goes through a concrete pump then it is 'good' concrete, whatever good concrete means, but I may say that Motts and MacAlpines took a great deal of comfort from this half-truth and indeed the power station still stands.

Dr Adam Neville Thank you. Who else?

Dr Pery Vassie, *Transport Research Laboratory* I have a comment for Nick Buenfeld. He said the use of the analytical approach to predict service life tended to produce conservative results because the diffusion coefficient of chloride ions in concrete decreased with time. The results of work at the Transport Research Laboratory (TRL) have shown that the surface concentration of chlorides tends to increase with time. The analytical approach, which does not take account of this effect, would therefore overestimate the service life. Our work has included a comparison of predicted service life and the service life obtained in practice; unfortunately the predictions were not sufficiently accurate to be useful.

In general the prediction of chloride concentrations at the level of the reinforcement is insufficient to assess the service life, which also depends on the threshold chloride concentration for corrosion and the rate of corrosion.

A practical method for ameliorating the effects of chlorides in concrete may be to use low water/cement ratio concrete where the electrical resistivity is sufficiently high to limit the corrosion current.

Dr Adam Neville Thank you very much. Can I add something to that based on my experience. There was a building with very thin reinforced concrete cladding. The inside was air conditioned and there was considerable carbonation of the concrete inside up to the steel. From outside, air-borne chlorides moved by rain advanced to the steel. So you get the worst of both worlds. But if I could just shift that to what Nick Buenfeld was saying about the ingress of chlorides, it is something that worries me because, as obviously you know, the chlorides are not fixed permanently. There is always an equilibrium situation of chloride ions in the pore solutions and therefore there are always chloride ions going into solution even if they had been fixed. It all depends on what other ions are there, so the situation is not of the kind where you can rely upon the fixings – I don't know whether you want to comment on that?

Dr Nick Buenfeld I think the importance of binding may be overlooked at first sight in a number of contexts. If there is bound chloride in concrete there will always be free chloride too. There are a number of processes that will release the bound chloride. One of these is carbonation, another is sulphate penetration. Just from that point of view one should perhaps be worrying more about total chloride than about free chloride. There is a widely-held view that I think our Chairman shares – I have read your recent papers – and a view that I certainly held until about nine months ago, which is that it is only the free chlorides that affect reinforcement corrosion and that if one is trying to define a critical chloride level it should be in terms of free chloride or chloride to hydroxyl ion ratio. We have recently surveyed the literature in great detail, almost a hundred papers, re-analysing the data and in several cases going back to the authors asking them for more information. We found that it is the total chloride that affects corrosion and the most appropriate way of expressing a critical value is chloride by weight of cement. The reason for this is, we believe, that much of the bound chloride at the surface of the steel is released prior to significant corrosion. I am really coming to my point now. Although binding slows down chloride penetration by taking it out of the pore water, binding also builds up the amount of chloride so that in low-chloride environments one can actually produce a critical case with a cement that binds a lot of chloride that could not occur with a cement that binds less.

Dr Adam Neville Thank you very much, Nick.

Mike Grantham, *MG Associates* I would like to widen the discussion on something a bit more fundamental related to all of these measurements that we are making, because we are making measurements on concrete and we are using that for predictive data to evaluate the condition of structures. I will talk about chloride measurements because it is something I am heavily involved in and it is a bit of a hobby horse of mine, but I recently went through an exercise of sending samples out to laboratories to see what variation we got in terms of chloride content on samples that were made of known composition.

This all started following some work where we obtained some data back from a laboratory where chloride gradients didn't look believable and we retrieved those samples and sent them out to a couple of other laboratories and got quite different results; in fact the two laboratories we sent them out to on the second occasion got quite close results to one another but totally different to the first lab. So we designed an experiment to evaluate just what kind of variability we have got between laboratories and we fabricated some samples of our own, disguised them as being from a car park with gradients and sent them out. The results were actually fairly horrifying; out of twelve laboratories that samples were sent to, five got results within $\pm$ 10% of the spec. values, the remainder were variously out from not very good to absolutely horrendous. Most of these were in fact NAMAS-accredited laboratories, I have to say, which suggests that the quality assurance procedures were not doing what they are supposed to do. NAMAS do have very stringent quality assurance procedures, so you are supposed to be testing samples of known composition throughout your measurements to ensure that what you are sending out is accurate. I can only surmise that that was not working properly in some of these laboratories.

So we are talking about chloride, but I do not think that is any different probably to any one of the other measurements that we make on concrete and we have got to be very, very careful in what we do, to make sure that the data we produce is actually accurate data and that can only be done by constant verification against samples of known composition or against controls to say what is happening. The other thing that is extremely important on this is that we actually deal with samples that are adequate for testing. One of the things I did on that exercise was to send out one of the samples deliberately very underweight, so it was unlikely to be representative of the concrete. The Concrete Society suggests for a pure dust sample 25 or 30 g would be a reasonable sample to take. As someone involved in testing for a number of years the number of times you get half a teaspoonful of dust sent to you for a chloride analysis is enormous, so on our exercise we sent out one sample that was underweight. Two labs noted it, one commented it was underweight, one reported it but did not comment at all, ten laboratories said nothing at all, so I think the testing authorities themselves have got to be professional in this and if they feel the samples are not adequate for testing, to tell their customers that this is the case. They are rather reluctant to do this, I must add, because they are afraid of their customers getting upset with them but it is something that is rather essential.

Dr Adam Neville Thank you.

Dr Pery Vassie At TRL we have had similar experiences to Mike Grantham. Large samples of powdered concrete from UK bridges have been thoroughly mixed and split to enable duplicate analyses to be carried out. On too many occasions the results of duplicate samples have been substantially different and could have resulted in money being wasted on unnecessary repairs or potentially serious corrosion being overlooked. This problem may arise from the use of electrical methods for determining the end point of the titration. Although these methods are allowed by BS 1881, the standard gives no advice on the procedure to be used. It could be that more frequent calibration is required to achieve satisfactory precision. There is a strong case for researching this problem.

Dr Adam Neville Well you referred there to the reliability basis of chloride tests. If I could comment in passing on the need for two separate tests. John Bungey pointed out the benefits of two non-destructive tests. The same situation exists in engineering design. You must obtain similar numbers by the two methods before you commit yourself. Then perhaps you could pretend you really have the same numbers. This is an engineering principle that you know, but there must be some more testing-people who want to say more.

John Glanville I think there are a number of factors to be considered in this and since Mike Grantham is our consultant and advises our chemistry lab. I am sure we get the results right. Fifteen years ago we used to be paid £12 to £15 for doing each chloride test; we are now able to get about £1.50 to £2. This puts extreme pressure on laboratories to short-circuit the system because they are not paid enough to do the job properly. They have to have the work to keep themselves busy and I am absolutely convinced that a large amount of chemical analysis and testing work is not done properly because the clients do not pay enough. The system of competition is so acute that you will only obtain routine testing work if your price is so low that is does not correctly reflect the cost of doing the work. We have refused to battle in that field and it has meant that we have lost a great deal of work and are still losing a great deal of work but those of you who work for public authorities and are forced to go out to tender, I think can expect to get dud results time and time again.

Dr Adam Neville Thank you.

A. Sandberg, *Messrs Sandberg, Consulting Engineers* Needless to say, John Glanville is absolutely right and of course there are only two laboratories to which you should be sending them in the first place. If you start going round for the cheapest, you are probably going to get what you deserve. You want to go to the charitable organizations like the two of us who do the job properly and lose a lot of money in the process!

My main point, however, was to say that we have all put much effort into improving diffusion coefficients and the properties of the various concretes; is there comparable research work going on on permissible crack widths, particularly with these new superconcretes?

Dr Nick Buenfeld Many concrete elements are uncracked, for example gravity structures, segmental tunnel linings and prestressed concrete elements. However, much reinforced concrete is inevitably cracked and the cracks act as a short circuit allowing aggressive species easy access through the cover and beyond. We are very interested in this and have recently been awarded an Engineering and Physical Sciences Research Council grant within the Materials for Better Construction Programme to investigate this issue, particularly with regard to the transport of water and chlorides.

Professor John Bungey I should like to respond very briefly to a couple of points that have been made – firstly, to Dr Vassie. As he knows, the question of standardization of electrochemical measurements of concrete fell just at the time when British Standards was shifting emphasis of new work towards Europe. Work did actually start on standards but unfortunately it is now out of date. We have been trying to review this with the co-operation of the Concrete Society and the Institute of Corrosion. We have made some progress but unfortunately, as you can tell, we are some distance from going into print.

The second point relates to what I think Mike Grantham was saying and, although based on the analogy of chloride measurement, a lot of what he has said is true in a broader sense. This relates to the need for more than one test. I am just reminded that in my talk I referred to the lack of awareness in this country of statistical methods for interpretation and for specification of numbers of tests required to achieve particular confidence limits. This applies to many types of test result. It is important that realistic allowances are made for the number of tests, the variability of the testing and the variability of the material when assessing any sort of test results. All the time, the values that we get should be considered in relation to what we could reasonably expect to get, and if they do not match then there is a good indication that something is wrong. Finally, of course, all these matters do rely very much upon having skilled staff carrying out the testing on site. It is certainly true of the sophisticated techniques and equally true of the more humble methods, including the rebound hammer which, though very simple, is easily abused and misinterpreted.

Dr Jonathan Wood I should like to endorse what has been said on the importance of statistical analysis. There is a good book by Mike Leeming's father (J.J. Leeming, *Statistical Method for Engineers*, Blackie 1969).

The problems of testing for chlorides are not just a laboratory problem; there are also major errors from insufficient sampling on site. When I was at Mott Macdonald we had reference cubes made up with different amounts of chloride added. These were sampled, as for the structure, and these reference samples were included with batches of site samples sent for test. This provided data on the variability arising from sampling and that from the laboratory performance. Laboratories giving erroneous results were not used again.

The high variability in chloride ingress in structures in the field makes it difficult to achieve a clear picture from which a basis for remedial and maintenance works can be derived. Sampling often has to be carried out within a short period of road closure. It is always better to collect a large number of small samples, to give more increments with depth and more sample locations, rather than use the large sample size stipulated in the BS. The local variations in ingress profiles, even between samples 100 mm apart, arise from local variations in the environment, variations in the local mix composition and variations in compaction during construction. We will need to obtain enough data to identify these effects separately if we are to analyse the data statistically to obtain a basis for calibrating durability design as George Somerville and I advocate. I believe that much of the data for this can come from investigations of deteriorating structures by organizations like STATS, rather than from laboratory studies.

Dr Adam Neville I think the discussion is getting almost nullified because I could comment that it depends what size you want to get; if you want to get the lower size you don't have to do all that. When you talk about flood, you are dealing with the distribution of extreme values and we don't wait for the effect of 5000 years of floods to establish the value. Anybody else?

Maurice Levitt, *private consultant* I am not selling anything, Chairman, apart from trying to sleep with a good conscience tonight and, as the title of this conference is Prediction Performance of Concrete, I just have a feeling we are beginning to drift a bit. Before I go on to that generality I should like to throw something at both John Glanville and John Bungey where I think there is a little bit of information missing. That concerns John Glanville's father, whose papers I read avidly when I joined the Precast Concrete Federation in 1956. I was told I had to spend the next 40 years of my life studying crazing of cast stone – a subject which I found singularly interesting at that moment but took a week or two to break that dream and I started looking into precast concrete and the durability aspects. The first durability aspect I looked into was permeability to water. The only test that existed then was a British Standard called 473/550 which was the British Standard then for plain and interlocking roofing tiles and W.H. Glanville developed the test which was a waxed cap which one stuck to the top of the tile and applied a Marriott bottle for 24 hours at a head of 200 mm (I think John will remember this one). W.H. Glanville was very good at sticking caps to concrete but knew nothing about the physics of the Marriott bottle because if that concrete was any good at all the Marriott bottle turned itself into a thumping great gas thermometer and increased the head by more than was the intention. But it was on that test that I based the ISAT, and although people blame me for the ISAT, the ISAT was a development of W.H. Glanville's original test which is applied both to precast and *in situ* concrete. The comment that John Bungey made about the ISAT and the BS 1881 pt 208 which is out for public comment was that we are still waiting for precision data. I think one should be very careful when one uses the word 'precision' when one talks about surface characteristics of concrete because if anybody observes concrete drying out after rain it always looks patchy, dark, medium dark and light in appearance and therefore I would suspect anybody who produced three ISATs nominally within 10% of each other over the same section of concrete was either cheating or being very, very lucky.

Now, back on to the generalities of this conference. There seem to be a couple of stools between which we are falling, Chairman, and what I would like to say is intended as guidance for people who have to design, or have to make concrete, or have to get their hands dirty; you either start off getting everything right and are not going to need any testing procedures thereafter – you can check if you have got everything right – or you start off not knowing you have got everything right, and then have to test it for performance. I think John Bungey hit one of the nails on the head when he was talking about resistivity and linear polarization methods because I think they have got a tremendous future. Thank you, Chairman.

Dr Kate Hollinshead, *Building Research Establishment* Following on from what the last speaker was saying, I would like to mention a project that we have been involved in on the Tees Barrage. This is a project funded by the Teesside Development Corporation and the barrage structure has a 120-year design life. After an initial four-year period the structure is going to be passed to another body and at that time a commuted sum of money will be handed over which reflects the amount of maintenance which it is considered this structure will need. What the Teesside Development Corporation has done is set in place a programme of monitoring which will assure them of the durability of the structure. The initial period of four years is rather short to demonstrate that the concrete will be durable for 120 years and so we are taking a mixture of approaches, somewhere between Jonathan Wood's destructive approach and John Bungey's non-destructive approach. We have one-metre cube blocks which are exposed to the same river environment as the barrage, from which we can take cores, do chloride analysis and measure permeability, and we have also corrosion probes embedded in parts of the structure and in the large blocks for comparison between the two. Even though the initial project is quite short I think that this is a way of providing information, in the first instance to the client to assure them of the basic durability of the structure, to assure the person to whom the structure is to be handed over and, of course, it is adding to the knowledge database that can be used in the future design of structures like this. I think we need more clients, who are aware of the possibility of this type of scheme, to come forward and show the foresight and provide a small amount of investment right at the beginning of the construction stage. This will help move knowledge forward by proving information from structures before things start to go wrong, and from concrete that is working in terms of durability. We have one scheme proceeding here, there are probably others, and I think the more the merrier.

Andrew Graham, *AEA Technology, Harwell* I see myself in the role of customer, the purchaser of the construction industry's products. In the nuclear industry we have used all types of reinforced concrete and other structures over 40 years or so and I think we have encountered all the problems with concrete durability that have been talked about today to a greater or lesser extent. These are real problems to us because they might affect safety and production that can cause real economic losses.

With hindsight, the construction industry has been able to explain and provide solutions for all these problems, but this event is timely in attempting to predict durability performance of new structures. In comparison with other products we buy as an industry the performance of structures is poor; not necessarily because the products are inadequate but because the perception is that they will last for ever. The claims made for the life of other components may be more modest, but are more generally achieved. I think there are two things that a customer looks for in a structure, realistic life time prediction and assurance of the product quality. You have shown today that these are inextricably linked.

The point made forcibly again today is that concrete is made of extremely variable components, made on site in conditions which are difficult to control and made by operatives who may have little training and experience. It is my view that these aspects need to be addressed rather than accepted, by improving training and controlling the conditions, so that it can be demonstrated that the assumed conditions have been achieved. This tends towards more use of factory production techniques.

Assurance of product quality is currently based on specification of materials and demonstration that at each stage of production the specified conditions have been met. Less attention is paid to testing of the finished product. How often are the key durability parameters such as cover, surface density, permeability and chemical composition checked *in situ* in the finished product? Are these finished parameters the ones that should be used for life prediction and acceptance criteria by the customer?

Dr Adam Neville Thank you. Has anybody any question to put to Peter Fookes? Well, could I? As I understood you, before I decide to use a particular aggregate I have to look back on the tectonic plate 2000 million years ago and then gradually find the history. Isn't there a slightly simpler way? When we buy cement, not only has the cement to conform to BS 12 or whatever but we can always ask for a test certificate which will tell us what are the specific properties and possibly even what is the range of specific properties of the cement shipment. Why is it that in order to get the answer we have to pay you every time rather than the quarries or the pits to be able to provide some information in the selling process of aggregates? I don't know how far you can go with this, but is there any scope in getting a sort of advance description of the sort of things we are talking about on top of the more traditional descriptions? Well, we are told something about flakiness which is relevant to permeability. If we are told something about the possibility of alkali–silica reaction under different circumstances, I know it would increase the cost but it would not increase the cost of the concrete in which you are going to use the aggregate because somehow or other the testing of aggregate should be done. Would you please inform us.

Professor Peter Fookes Thank you, Mr Chairman. About 20 years ago, when I was looking at a working aggregate quarry for either Sandberg or STATS, I do not recall whom (I have been a consultant for STATS since the mid-seventies, and a consultant for Messrs Sandberg since the mid-sixties, and it is rather nice to mention them here!), I asked the quarry management had they any test certificates for their product and they produced from a drawer a rather scrubby sheet of paper which originated from a well-known laboratory in 1938, giving test results on just one sample. When I queried this, they said 'Well everybody else has been satisfied with this for the last 30 or 40 years'! Times have changed, of course, but the development of systematic study of engineering geological aspects of natural aggregates and aggregate sources is still, I believe, one of the last, if not the last, bastions to be scaled.

To answer you more formally, Adam, I believe that there are three levels of aggregate–quarry investigation to consider.

Firstly, if you are a big quarrying organization you probably have professional geologists in the woodwork, with plenty of experience, and you probably have a good quality control system in place and you are quite familiar with your quarry and your product because you have made a systematic study of it, both from the geological and production viewpoints. By and large you have probably been in the business long enough for there to be a track record of your aggregate in use and, under such circumstances, there is no need to make a fresh appraisal each time a buyer orders aggregate from your quarry. I see that Richard Fox is in the audience – his organization, I understand, has just such a system in place. I hope that future developments of the standards and codes will require that large quarries will audit themselves, perhaps with occasional outside checks, and produce regular QA/QC certificates.

Secondly, smaller quarry organizations, without a qualified geologist(s), I believe should be required in the future to instigate an initial professional appraisal and occasional reviews by an experienced geologist. You only have to look into any working quarry to see that the rock (or sand/gravel pit) does vary naturally from slightly to quite strongly on a blast-by-blast basis: such variation is not always easy to spot, even with experienced professional eyes (e.g. figures 6.9(a) and (b) of my paper). Such quarries should carry out routine testing for QA/QC certificates in addition to the occasional regular geological reviews. However, this situation, I repeat, is for some time in the future.

Thirdly, I consider a greenfield site needs initially a systematic study along the lines of Table 6.5 in my paper followed by an assessment later of the aggregate qualities along the lines of Table 6.6. I believe that, depending on the size of the quarry and the organization running it, the QA/QC procedures should be those I have outlined above for a large or a small quarry.

There are risks in quickly dropping into a quarry and making a more or less off-the-cuff evaluation – as I am often asked to do. Some geological nasty minerals for concrete (e.g. opal) may not be visible in a quarry face on any one particular day. Routine testing may never pick up such nasties. Observation and geological acumen should. A careful, systematic study, preferably supported by experience of the aggregate from the particular quarry over a period of years, I believe, is the best practice.

Now a plug: if you have not bought it already, I would recommend you buy the second edition of the Working Party Report on Aggregates, by the Geological Society – no doubt available for sale in the office here today!

R.H.R. Douglas, *consultant to Sir Frederick Snow & Partners Ltd* I hesitate to speak here amongst this number of specialists – I am certainly not a specialist myself in concrete, but on the last point that was made I do have much more to do with Marshall mix asphalt for use in runway pavements and for a mix of that nature, where the quality of the aggregate is extremely important, as it is indeed in concrete, we certainly identify which quarry stone is coming from and how it mixes together so we have no problem in getting our information. I see no reason why it should not be possible to get the same information as far as concrete is concerned.

I would like to make another point also and that is, of course, we are all looking to the future, we tend not to go back and look at what has happened; there are a lot of structures, highway structures, which were built earlier in the century and which are currently being demolished to make way for wider roads. For example, on one of our North Circular Road projects we have just demolished a reinforced concrete viaduct. Walking round that site I have to say I see very little sign of corrosion in the reinforcement there. I wonder how much attention is being given, should be given, to examining these structures when they are being removed to see what information can be got from them. Not all of them are disasters. Many of them are showing very little sign of deterioration and I do not think this has been mentioned during the course of today.

Dr Adam Neville Thank you very much. If you look at it, the programme provides for the Chairman's summing up. You may have noticed that I have indeed been making copious notes and therefore, as a good and experienced chairman does, I prepared my summing up before I was likely to be confused by what you are going to say. I am not joking. You will be able to re-live the presentations, partly because you have them in the papers, partly because the book will appear – Nick Clarke assures us, beautifully produced – and the book will include not only the papers but also the discussion which I think should be truly illuminating. But what I would like to do, if you will allow me, is to go through ten points which I feel are of particular importance.

Some of them were mentioned to a greater or lesser extent, some were not. They are not in an order of priority. I think the first one was not mentioned, which is that we need the client or his representative to establish his requirements with respect to service life as well as maintainability. By putting the two in a coupled way, because for example we should know if the client wishes to have regular maintenance which is not done in buildings, this means expenditure to which no thought is given in advance. Or, does he want no interference with his operation of the structure? If so, he must spend more money up front and to take this decision, the client must become fully conversant with life cycle costing. Only bits of this are known but I am trying to put the whole thing together to be put to the client up front.

My second point puts a gloss on the previous one, something which has been mentioned by many people, and that is that the old propaganda about concrete structures being maintenance-free must be stopped and abandoned completely.

My third point concerns the design life. Various figures were thrown out today – I think the maximum bid I heard was 300 years. I think the life must be reasoned, reasoned properly, so that we do not move to the extreme of an excessive length of service life. My argument, and it is not an original argument, is that social and functional criteria are paramount. Just to give a pungent example: 100 years ago to build a privy with 100 years' service life would have been unwise.

My fourth point is one which I have already made in my introduction, but I think it is a point which bears repeating: we should ensure that the client recognizes durability as **a** primary design requirement but not as **the** primary design requirement. In other words, I don't think we want to shift from one extreme to the other. Usually, there is an absence of durability requirements in established documents. As far as I can tell there is almost nothing in the Building Regulations; in the codes the advice is very poorly quantified. There are no accepted design procedures which include an explicit and probabilistic approach based on service life.

Well, my fifth point is that structural design and detailing sometimes unnecessarily contribute to poor durability. The example which I had in mind I think was shown by George Somerville but I did not see the slide well enough. My example is from Norway, a country which has many more jetties than people. It was found in an inspection of those jetties that beam and slab construction led to much more corrosion than flat-slab construction because of moisture conditions in corners and re-entrance areas underneath the jetty above water. Well, this brings me to my sixth point, the need for a proper description of the microclimate in the various parts of the structure and not just a description of the environment of the whole structure. I don't think that point has been considered much except insofar as Nick Buenfeld spoke about corrosion of a sheltered and unsheltered area. It is well known that, with respect to carbonation, gable ends do better than walls underneath the eaves, but in more important structures I have seen enormous differences and I have not seen proper mention of them. The extreme example which I think Peter Fookes and I shared was the deterioration of concrete around fire hydrants where sea water was used. The wetting was quick and the drying was prolonged. The imbibition of water was far greater than tide zone, wave zone, splash zone or anything.

Well, my seventh point is the need for a better knowledge of the behaviour of materials under different conditions. I hope we all accept now that the days of pure Portland cement are over, that we are now going to make concrete with a minimum of two and probably three or four what I call cementitious materials (different people use different names) and of course admixtures. Well, one example which I came across concerns carbonation when a cementitious material includes slag or fly ash. Tests were done but were not helped by research of the kind where concretes with and without fly ash were compared, but wet curing lasted one day. You don't have to have a Ph.D. to figure out that that sort of curing doesn't contribute much to the microstructure of a cement-hardened paste. Unless somebody reads a paper like that very carefully (and by very carefully I mean with some advanced knowledge of what can happen and what should happen) it is difficult to draw proper conclusions.

My eighth point is very different. It is to do with testing methods. We heard about at least four mechanisms through which water can ingress into concrete but I have seen cases where the test method described is that of permeability under a head of water while the actual mechanism involves no head at all. Testing the performance of a particular concrete was done by way of a comparison of two or more concretes. This was done by measuring permeability because there exists a German DIN 1045 test. So it is very easy to refer to it, while if you want to do a wick test or a sorption test it is not a standard test. I think engineers are, we are, given to using tests which give numbers just like that. There was reference made to the T259 test for chloride permeability. This test does not measure chloride permeability, it measures an electrical current but you get a number and you put a ring round it and use a highlighter and say 'that is the outcome'. I think we are to blame in liking those sorts of tests and sticking to them in our specifications.

Well, my ninth point, and I reach my hobby horse, is improved formation and training of all those involved in concrete. There were a few remarks made now and again about the unskilled labour force involved in concreting. On one major project, as we ascertained after the event, there wasn't a single person on the job who had experience of that type of construction.

My tenth point is really commercial. I am doing it very, very subtly which, as you must have noticed, is my traditional approach. A successful implementation of a selection of ingredients and proportions in concrete requires experience coupled with the knowledge of the influence of various factors upon the properties of concrete. This knowledge must be based on an understanding of the behaviour of concrete. This is given in my new edition of *Properties of Concrete: experience, knowledge and understanding*. I think this meeting today, of people with experience and with knowledge, I am sure has furthered our understanding of what happens in concrete and this is the way forward.

All that falls to me now is to thank a number of people. First of all, to thank John Glanville for inviting us, for thinking of this occasion, for organizing it and for seeing it through and for giving us a useful and pleasant time. I would like to thank the speakers; I would also like to thank the discussers and I would like to thank all attendees, for without all of you we could not have had this conference. I think thanks are also due to those who organized this excellent event. John Glanville was generous in inviting all of us from the concrete world. He was also kind to invite students from Queen Mary and Westfield College, this year's students and last year's students from the M.Sc. course in Geomaterials. Finally, happy birthday to STATS!

Keyword index

This index has been compiled from the keywords assigned to the papers, edited and extended as appropriate. The page references are to the first page of the relevant paper.